SpringerBriefs in Electrical and Computer Engineering

Speech Technology

Series Editor
Amy Neustein

For further volumes:
http://www.springer.com/series/10043

Editor's Note

The authors of this series have been hand selected. They comprise some of the most outstanding scientists—drawn from academia and private industry—whose research is marked by its novelty, applicability, and practicality in providing broad-based speech solutions. The Springer Briefs in Speech Technology series provides the latest findings in speech technology gleaned from comprehensive literature reviews and *empirical investigations* that are performed in both laboratory and *real life* settings. Some of the topics covered in this series include the presentation of real life commercial deployment of spoken dialog systems, contemporary methods of speech parameterization, developments in information security for automated speech, forensic speaker recognition, use of sophisticated speech analytics in call centers, and an exploration of new methods of soft computing for improving human–computer interaction. Those in academia, the private sector, the self service industry, law enforcement, and government intelligence are among the principal audience for this series, which is designed to serve as an important and essential reference guide for speech developers, system designers, speech engineers, linguists, and others. In particular, a major audience of readers will consist of researchers and technical experts in the automated call center industry where speech processing is a key component to the functioning of customer care contact centers.

Amy Neustein, Ph.D., serves as editor in chief of the *International Journal of Speech Technology* (Springer). She edited the recently published book *Advances in Speech Recognition: Mobile Environments, Call Centers and Clinics* (Springer 2010), and serves as quest columnist on speech processing for Womensenews. Dr. Neustein is the founder and CEO of Linguistic Technology Systems, a NJ-based think tank for intelligent design of advanced natural language-based emotion detection software to improve human response in monitoring recorded conversations of terror suspects and helpline calls.

Dr. Neustein's work appears in the peer review literature and in industry and mass media publications. Her academic books, which cover a range of political, social, and legal topics, have been cited in the Chronicles of Higher Education and have won her a pro Humanitate Literary Award. She serves on the visiting faculty of the National Judicial College and as a plenary speaker at conferences in artificial intelligence and computing. Dr. Neustein is a member of MIR (machine intelligence research) Labs, which does advanced work in computer technology to assist underdeveloped countries in improving their ability to cope with famine, disease/illness, and political and social affliction. She is a founding member of the New York City Speech Processing Consortium, a newly formed group of NY-based companies, publishing houses, and researchers dedicated to advancing speech technology research and development.

Oi Yee Kwong

New Perspectives on Computational and Cognitive Strategies for Word Sense Disambiguation

 Springer

Oi Yee Kwong
Department of Chinese, Translation and Linguistics
City University of Hong Kong
Kowloon
Hong Kong

ISSN 2191-8112   ISSN 2191-8120   (electronic)
ISBN 978-1-4614-1319-6   ISBN 978-1-4614-1320-2   (eBook)
DOI 10.1007/978-1-4614-1320-2
Springer New York Heidelberg Dordrecht London

Library of Congress Control Number: 2012938040

Printed on acid-free paper

Springer is part of Springer Science+Business Media (www.springer.com)

To my parents

Preface

The importance of word sense disambiguation (WSD) has been well recognised in computational linguistics since the inception of the discipline. Throughout these years, despite the lack of strongly convincing evidence or absolutely clear findings to the contrary, there is always a conviction that WSD is beneficial to practical natural language applications, typically including machine translation and information retrieval. This discrepancy will continue to exist as long as the state of the art has not reached the threshold which is good enough to guarantee the reliable deployment of some WSD component in real applications, or to ensure that notwithstanding the errors, the performance of the latter will be boosted rather than deteriorated. Recent evaluation exercises, in particular SENSEVAL and SEME-VAL, have revealed a lot about the nature and requirements of WSD. It is almost beyond any questions now that multiple knowledge sources (syntactic, semantic, and pragmatic information inclusive) are needed for effective disambiguation; and there is more or less a consensus these days that automatic WSD should content itself with homograph distinction. However, when we observe that even the best systems show differential performance on different target words, and systems using the same set of knowledge sources do not always make the same predictions, we cannot help asking how thoroughly the lexical sensitivity of the task has been understood and taken into account. Computational linguists have focused heavily, if not entirely, on the technical aspects of the systems, and few had attempted to give any serious or extensive account of the cognitive aspects of WSD. While individual systems could be fine-tuned by engineering the learning algorithms and the feature sets, the obstacle to overcoming the currently plateaued performance obtained from supervised systems may rest with some intrinsic properties of words and senses closely related to our cognition. In fact, if we look at the history of WSD, which is more than half a century now, computational linguists and psycholinguists had backed up each other in the early days, but they had then diverted from each other, more and more. It is thus time to take a step back, to re-examine the computational strategies (by machines) and the cognitive strategies (by humans) for WSD in parallel, and to explore alternative classification of senses which might shed light on their differential information susceptibility. By doing so, a research agenda is proposed for a closer look into the lexical sensitivity and thus the different information demand for WSD. Some of the ideas in this book have already sprouted during my doctoral studies, for which I am always indebted to the late Prof. Karen Sparck Jones who had patiently guided me through those years. Although these ideas have remained dormant for some time, I hope their revitalisation through this book will lead to their blossoming and fruiting before too long.

Hong Kong, February 2012   Oi Yee Kwong

Acknowledgments

The work reported in this book was partially supported by grants from the Research Grants Council of the Hong Kong Special Administrative Region, China (Project No. CityU 1508/06H), and the Department of Chinese, Translation and Linguistics of the City University of Hong Kong. I am grateful to Dr. Amy Neustein, the Series Editor, for her invitation and comments to the earlier drafts of this Brief.

Contents

Chapter 1
Word Senses and Problem Definition

Abstract This book is about word sense disambiguation, the process of figuring out word meanings in a discourse which is an essential task in natural language processing. Computational linguists' efforts over several decades have led to an apparently plateaued performance in state-of-the-art systems, but considerable unknowns regarding the lexical sensitivity of the task still remain. We propose to address this issue through a better synergy between the computational and cognitive paradigms, which had once closely supported and mutually advanced each other. We start off with an introduction to the word sense disambiguation problem and the notion of word senses in this chapter. While the psychological reality of word senses is beyond doubt, the boundaries between senses could be fuzzy. We discuss various models for representing senses and suggest that the discreteness assumption held by most mainstream systems is relevant to the perception of word senses rather than their definition.

1.1 Problem Definition

It so happens that our communication, in different languages alike, allows the same word form to be used to mean different things in individual communicative transactions. The consequence is that one has to figure out, in a particular transaction, the intended meaning of a given word among its potentially associated senses. While the *ambiguities* arising from such multiple form-meaning associations are at the lexical level, they often have to be resolved by means of a larger context from the discourse embedding the word. Hence the different senses of the word "service" could only be told apart if one could look beyond the word itself, as in contrasting "the player's service at Wimbledon" with "the waiter's service in Sheraton". This process of identifying word meanings in a discourse is generally known as *word sense disambiguation* (WSD).

O. Y. Kwong, *New Perspectives on Computational and Cognitive Strategies for Word Sense Disambiguation*, SpringerBriefs in Speech Technology, DOI: 10.1007/978-1-4614-1320-2_1, © The Author(s) 2013

Research on automatic WSD has enjoyed a history as long as computational linguistics itself. Back in the 1950s when machine translation research emerged as ambitious projects, the importance of disambiguating word senses has already been realised since word sense ambiguities often surface as translation differences. A typically cited example is "duty" in English should be translated to "devoir" or "droit" in French depending on whether the word is used in its "obligation" or "tax" sense respectively (e.g. Gale et al. 1992a). The conviction that some sort of WSD is needed for nearly every natural language processing (NLP) application is perhaps what has kept WSD research active all along. According to Hurford and Heasley (1983), "the sense of an expression is its place in a system of semantic relationships with other expressions in the language". Automatic WSD thus largely depends on the knowledge sources capturing the various semantic (and possibly other) relations among words available to a system, and subsequently its ability to uncover and deploy these relations among words in a text.

The common practice of current mainstream WSD is to treat the task as one of classification. Systems thus attempt to assign the most appropriate sense among those given in a particular sense inventory, typically some dictionary or lexical resource, to individual words in a text. The principle is similar to part-of-speech (POS) tagging in which a tagger labels each word in a text with the most appropriate POS tag from a fixed tagset. Hence the process is sometimes known as sense tagging (Stevenson 2003).[1] The use of predetermined senses in a given sense inventory relies on two basic assumptions, which are nevertheless debatable. First, senses exist and can always be represented in some way. Second, senses are discrete, and in principle they can be unambiguously assigned to novel occurrences of words in a discourse. We will return to the controversy of these fundamental points in the next section.

Notwithstanding its long research history and the many encouraging results reported from state-of-the-art systems, WSD is still receiving much attention these days, though probably less as a task in itself but more for its contribution to real language processing applications like machine translation and information retrieval. In fact, WSD has often been considered a means rather than an end right from its birth, and as an enabling technology for larger NLP applications. However, in vivo evaluation of WSD, unlike its in vitro counterpart, apparently brings more disappointment than expectation (e.g. Sanderson 1994; Voorhees 1999). Errors in WSD often adversely affect the application. Resnik (2006) suggested that the traditional conception of WSD via an explicit sense inventory may not be practically appropriate for some NLP applications, and many others (e.g. Ide and Wilks 2006; McCarthy 2006) questioned the suitability of the sense granularity in conventional WSD research for real applications.

---

[1] Stevenson (2003) distinguishes sense tagging from sense disambiguation, where the former attempts to disambiguate all words in a text but the latter only does so for a restricted set of words. In both cases, the annotations applied are senses from some lexicon.

In addition to the issue of sense granularity, the *lexical sensitivity* of WSD has also long been realised (e.g. Resnik and Yarowsky 1999). Investigating the effects and contributions of various kinds of knowledge sources on disambiguation has been held as important research directions (e.g. Ide and Veronis 1998), but apparently the issue still remains an issue after more than a decade. Work on WSD by computational linguists has primarily focused on the technical aspects. Recent SENSEVAL and SEMEVAL evaluation exercises[2] have shown that state-of-the-art WSD systems tend to rely on an optimal combination of disambiguating features within some machine learning framework and ensembles of a variety of such trained classifiers (e.g. Mihalcea et al. 2004), but even the best systems show differential performance on different words (e.g. Pedersen 2002). Have we really reached an insurmountable plateau of about 70% accuracy reported by contemporary systems? Have we taken the lexical sensitivity of WSD seriously enough? The objective of this book is therefore to take a step back to see what have been learned so far from WSD evaluation and to look further into the nature of different target words, that is, the words to be disambiguated. In particular, we will (re-)examine in parallel the computational strategies and the cognitive strategies for WSD used by machines and humans respectively. We believe such a reunion of the two paradigms, which were once closely linked especially in the early days of WSD research, will illuminate us with further insight to enable a better understanding of the lexical sensitivity of the task.

This introductory chapter starts with a discussion on the notion of word senses. We will look at the various views on its reality, discreteness and representation, and see how the aforementioned assumptions on word senses have formulated the problem of WSD in most current research.

1.2 The Notion of Word Senses

Do word senses exist? It sounds irrational to ask this question given that we are working on word sense disambiguation which by itself already presupposes the existence of word senses. Are they not simply the meanings of words one finds in most dictionaries? But many people are really sceptical about it, including experienced lexicographers who work with senses for their career. For example, Kilgarriff (1997) had his paper remarkably entitled "I don't believe in word senses", quoting from Sue Atkins' response to a discussion which assumed discrete and disjoint word senses, and suggested that word senses exist only relative to a particular purpose or task. Hanks (2000) also questioned whether word meanings exist and proposed a "Yes, but ..." answer to it, arguing that they are

---

[2] SENSEVAL and SEMEVAL provide a common platform for evaluating WSD systems, where participating systems submit their results on the same test data and are evaluated against the same gold standard. WSD evaluation will be further discussed in Chap. 3.

misleadingly represented by traditional dictionary definitions. Words are more properly considered to have meaning potentials made up of many components, which are only activated in context.

The controversy lies in the difficulty to pin down a precise definition of sense. Sense is an abstraction. We cannot possibly point to something and say that is the sense of a word. We can only grasp the sense of a word via our understanding of the meaning of a message where the word is used to express it.

As a linguistic concept, word sense covers only the linguistic (or semantic) meaning of words, in contrast to other types such as speaker (or pragmatic) meaning. For example, imagine this scenario: I am taking a stroll with a friend while he complains about someone's misconduct. Seeing a dog coming towards us, I say to my friend: "What a hound!" At this moment, my friend would probably think: Am I referring to a four-legged animal or a morally reprehensible person? Am I asking him to protect me or am I showing my stand about that person's behaviour? As far as the *sense* of the word "hound" is concerned, only the first question is relevant. To disambiguate word senses is therefore to find out the *use*, among all possible *uses*, that a word adopts at a particular occasion. Although the distinction is a linguistic one, the process does not preclude the use of extralinguistic information. For instance, in order to get my message, my friend might need to know my temperament and if I am usually afraid of dogs.

Word sense ambiguity arises when the same word can potentially be used to mean different things. Let us not trace the historical reason why language has evolved this way, but simply accept the presence of ambiguity as a linguistic fact. Linguists used to distinguish between two kinds of word sense ambiguity: *homonymy* and *polysemy*.[3] Homonymy refers to a word having meanings which lack obvious semantic relations among themselves, normally because they have distinct etymological sources (e.g. "pen" as "a writing instrument" or "an enclosure for livestock"). Polysemy, on the contrary, refers to a word having different but closely related senses (e.g. "pen" as "an enclosure for livestock" or "an enclosure in which babies play"). It can be systematic or idiosyncratic, and sense shifts are not uncommon (Clark and Clark 1979; Copestake and Briscoe 1995). For example, there is regular polysemy, such as words for animals like "chicken" and "lamb" can systematically refer to either the animal itself or the meat of it being consumed. In the case of metonymy, "ham sandwich" could be extended to refer to the person who has ordered the food. Given the close relations among the senses, polysemy is often considered more problematic for WSD. The variation in the distance among senses also poses a problem to where one should draw the line between genuine polysemy and what might just be the *vagueness* of sense (e.g. Lytinen 1988).

---

[3] Homonymy and polysemy may occur within or across syntactic word classes, and for the latter they are more often known as categorial ambiguities. Since POS tagging would have taken care of most categorial ambiguities, our main concern for WSD will be the ambiguities within the same POS.

Although most people may not consciously think in terms of word senses when they speak or write, there is psychological evidence to support the reality of word senses, as human language users are found to have the ability to distinguish different meanings carried by the same word used in different communicative transactions (e.g. Jorgensen 1990). Nevertheless, considerable debate has persisted on how hard (or soft) a boundary one could draw between two meanings of a word, which Aitchison (2003) discussed as the fixed-fuzziness issue.

Up to this point, the readers might have noticed that we have not tried to distinguish between "sense" and "meaning". But are they really equivalent? Philosophers (and interestingly, many lexicographers alike) will probably say no. Our daily usage of the two words might have hinted on their difference. Imagine a person makes a statement and another responds as follows:

A: John is generous.
B: In what sense? / What do you mean?
A: ...

"In what sense" is apparently like a multiple-choice question, assuming a finite set of generally accepted scenarios in which someone can be qualified as "generous" (e.g. John is "kind and forgiving" or "liberal in giving"); whereas "what do you mean" sounds more like an open-ended question, allowing more flexible interpretation and clarification by Speaker A according to the situation (e.g. John has donated a large sum of money to charity or he has sponsored a lot of presents for the Christmas party).

The "meaning" of a word is therefore the message intended by the speaker to be expressed by it, and the specification of the message might vary, however slightly, according to the linguistic and extralinguistic context of the particular transaction. The "senses" of a word, on the other hand, is like a snapshot of all its possible meanings, frozen at a particular time, and captured with a particular resolution. Therefore, strictly speaking, meanings associated with every single use of a word are continuous but senses (as listed in a dictionary) enumerating the *core* meanings are discrete. As remarked by Kilgarriff (2006), lexicographers attempt to give a full account of the words of a language by abstracting distinct meanings from usage patterns sharing sufficient commonality from corpora data and describing them in the form of dictionary senses, mostly via definitions. The task is hard since the clustering process is not always straightforward. There is often no decisive way to identify where one sense of a word ends and its next sense begins.

It then turns out that if the word senses described in a dictionary or other finite inventories are not exactly the meanings that words will carry in individual communicative transactions, the whole paradigm of automatic WSD as it is practised today will be questionable. However, computational linguists are apparently happy with what they are doing and achieving in WSD. As Agirre and Edmonds (2006) noted:

"Concerns about the theoretical, linguistic, or psychological reality of word senses notwithstanding, the field of WSD has successfully established itself by largely ignoring
them, much as lexicographers do in order to produce dictionaries." (p. 9)

As Wilks et al. (1989) suggested, we need some representation of lexical ambiguity anyhow to model the handling of "ambiguous words" computationally.
Hence, to computational linguists, it is the clarity, consistency, and coverage of the
range of meaning distinctions which they care more.

How then should we compromise with the assumption of discrete senses? Let us
turn for a while to the various models of representing word senses in the next
section, before we return to this issue in Sect. 1.4.

1.3 Representation of Word Meaning

Despite its fuzziness, there is nevertheless the need to encode the meaning of a
word with other words in the language. However imprecise such representations
might be, they serve the purpose for language users to "make sense" of a word
when it is heard or used. We summarise and compare three main models for
defining and representing word senses, namely the *descriptive model*, the *relational model*, and the *generative model*, with respect to how they view the stability
of senses and how explicitly they define senses. By *explicitness*, we refer to the
preciseness of the relationship between the word being defined and the word(s)
used to define it. This should not be confused with *concreteness*, which we refer to
as the perceptibility of any image that a definition carries.

1.3.1 Descriptive Model

The descriptive model is the most intuitive and common model for defining senses,
which is found in most conventional monolingual dictionaries. Senses are usually
defined via some necessary and sufficient conditions, similar to constructing a
prototype of a concept. A typical example is to define "bachelor" as "an
unmarried man", in which HUMAN, MALE, ADULT, and UNMARRIED are the
necessary conditions.

A major drawback of this model is that definitions can be prone to inconsis-
tency and inadequacy. Nonetheless, there is rich information in dictionary definitions, often more than just linguistic information. Although the relations between
the headword and the words in its definitions are implicit, the patterns of definitions are often exploited to make the relations explicit for computational use (e.g.
Amsler 1981), which will be further discussed in Chap. 2.

The style of definitions can vary from dictionary to dictionary, depending on
their individual editorial philosophy as well as their target audience. For example,

the Longman Dictionary of Contemporary English (LDOCE) restricts the words in its definitions to a defining vocabulary of about 2,000 words and uses a rather regular syntax for the definitions (Procter 1978). As a pioneer of corpus-based lexicography, the Collins COBUILD English Dictionary defines words by explaining their uses with authentic examples from naturally occurring English texts (in contrast to artificial examples as used in some other dictionaries) in the form of full-sentence definitions (Hanks 1987; Sinclair 1987). It is also well known that dictionaries differ from one another in terms of sense granularity. For instance, although the Cambridge International Dictionary of English (CIDE) resembles LDOCE in almost every way, it is claimed to have more fine-grained senses than LDOCE (Harley and Glennon 1997).

The descriptive model is biased towards a static view of senses, which suggests that senses are discrete and enumerable. This appears reasonable on the assumption that each word has a set of core senses and they are listed in the dictionary. However, the model is often criticised for failing to cover word senses which are not core, and reducing word senses to decontextualised definitions. Despite this, the descriptive model remains a common model for general-purpose sense representation.

1.3.2 Relational Model

Relational models define word senses directly in terms of their semantic relations (e.g. synonymy, antonymy, hyponymy, etc.) with other words in the language, in which the impact of connectionist models for the mental lexicon proposed by psychologists (e.g. Collins and Loftus 1975; McClelland and Rumelhart 1981) is most evident. Here we describe two specific models of this kind.[4]

Sparck Jones (1986) argues that senses can only be most appropriately defined in terms of synonymy. She defines a sense in the form of a "row", which consists of words substitutable for one another at a particular position in a sentence without changing the ploy of the sentence. Rows are bootstrapped from definitions with synonyms in a dictionary and substitutability is tested against the usage illustrations in the dictionary. For example, 26 rows were obtained for "action" from the entries in the Oxford English Dictionary, such as "action gesture motion movement", "action process suit case", etc. Rows can be combined into more general classes to form a semantic structure for the words in the language, which constitutes a bottom–up model of constructing thesaural classes. The classification of word senses in WordNet, one of the most widely used computational lexical resources in WSD to date, is also grounded on synonymy (Miller et al. 1990).

Relational models also allow senses to be defined by other forms of word clusters (e.g. Schütze 1992; Kilgarriff 1997). Schütze's (1992) Word Space Model

---

[4] See for example Evens (1988) for a fuller account of relational models.

defines senses as context groups, derived from co-occurring words in a corpus and represented in a multi-dimensional vector space. For example, with "legal" and "clothes" forming two dimensions, word vectors for "judge" and "robe" can be represented based on their co-occurrence counts with the two dimensions. A context vector is given by the centroid of the word vectors occurring in the context. The context vectors of an ambiguous word are clustered into a predetermined number of groups, and the various senses are represented by the centroids of these context groups.

Relational models also view word senses as a finite set, except that they are often entirely data-driven, and it is possible that some intuitively core senses are excluded. However, relational models differ in their explicitness. For instance, Sparck Jones' rows are always governed by synonymy. Other clustering methods may be less explicit since the relations which tie up a cluster are not precisely known.

1.3.3 Generative Model

Pustejovsky (1991) proposed a dynamic model for representing word senses, where the encoding is actually left open-ended. In his generative model, compositionality is assumed, and lexical meaning is captured by various levels of representation, namely "argument structure", "event structure", "qualia structure", and "lexical inheritance structure". The various levels can be connected via type coercion. Ambiguity resolution therefore becomes a process of highlighting mutually compatible and relevant lexical components of words and phrases, achieved by "co-composition". For example, by predefining the general behaviour of a word like "fast" as always predicating of an event type (which matches the telic role of a nominal), subtly different uses of the word can be encoded. Consider the meaning of "a fast car". Given the following qualia structure for "car":

$$\begin{bmatrix} car(x) \\ CONST = \{body, engine, \ldots\} \\ FORMAL = physobj(x) \\ TELIC = drive(P, y, x) \\ AGENTIVE = artifact(x) \end{bmatrix}$$

the adjective "fast" can be resolved as modifying the event "drive", which means the process but not the car itself is fast. Similarly, given the telic role of "motorway" as follows:

$$[Telic : travel(P, cars)\hat{\ }on(P, x)]$$

"fast" in "a fast motorway" should describe "the process of travelling on a motorway", rather than the motorway itself (Pustejovsky and Boguraev 1993). It is otherwise impossible to list such slight sense variations exhaustively. Such a

Table 1.1 Models of sense representation

Model	Sense stability	Sense relations
Descriptive	Static	Implicit
Relational	Static (Data-driven)	Implicit/explicit
Generative	Dynamic	Explicit

compositional approach for sense representation is echoed by lexicographers and psycholinguists (e.g. Hanks 2000; Pinker 2008).

Relations between words are explicitly encoded in the generative model, although the relations are more logical than linguistic. In addition, the model has been shown to be able to represent systematically some regular polysemy and sense extensions[5] (Copestake and Briscoe 1995). Despite its advantages, the model makes a constrained form of dictionary which is tricky to be grounded properly in the first place, thus limiting its utility and popularity in WSD.

1.4 Softening the Discreteness Assumption

The sense representation models have demonstrated that word senses can be represented in some way. Table 1.1 summarises the various models in terms of their views on the stability of senses and the explicitness of sense relations. Computational treatment of WSD often considers it a classification task, which requires a finite set of categories to start with. Hence the sense inventories used in current mainstream WSD research are often based on resources following the descriptive or relational models.

However, the assumption for discrete senses encompasses two questions: Will senses ever be the same so that a sense previously seen can be used to name a later one? Are sense boundaries definite enough so that we can say a new occurrence of a word falls under one sense but not the other?

As mentioned in Sect. 1.2, if the meaning of a word is tied to its individual uses, and no two occasions where the word is used (including linguistic and extralinguistic contexts) will ever be identical in the strictest sense, it follows that a finite set of predetermined senses will never be sufficient for labelling new uses. But people do not use words randomly. They use a word because they believe the word will convey a meaning which has been conveyed by it before (except when people deliberately use words in creative ways). Thus, although there are two occasions of use, we can make a compromise by saying the two occurrences carry the same sense if they appear in contexts which are similar enough to justify that they most probably convey the same message.

---

[5]   e.g. logical metonymy (a fast car, a fast typist, a fast motorway, etc.), broadening (e.g. a cloud of something), nominal metonymy (e.g. the ham sandwich), portioning (e.g. three beers = three portions of beer), grinding (e.g. lamb as the animal or the meat), and so on.

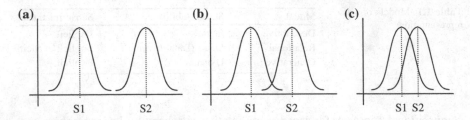

Fig. 1.1 Perceptual discreteness and boundaries of word senses

We can think of the situation symbolically as in Fig. 1.1. A sense inventory only lists a fixed number of senses for a word (e.g. S1, S2), but there is a range of contexts (the span covered by each curve) within which we will perceive a novel use as the listed sense. The discreteness of sense held in the second assumption should therefore be more appropriately viewed as a perceptual kind, over a continuum of contexts and uses. Hence it is our perception which is made discrete, but word sense itself is not necessarily so. This has no contradiction to other viewpoints on the discreteness of sense, such as Lyons' (1981):

"Discreteness in language is a property of form, not meaning." (p. 148)

The classic "Bank Model", which was once refuted by Kilgarriff (1993) largely from a lexicographer's point of view but later defended by Wilks (1998) mainly from a computational linguist's perspective, can then be considered to fall on one extreme of the continuum depicted in Fig. 1.1.[6] In other words, it only covers homonymous cases in which two senses have little semantic relation and are thus often found in very distinct contexts, as represented by Fig. 1.1a. For the rest of the continuum, there are different degrees of polysemy where senses are more closely related, as in Fig. 1.1b and c. Therefore senses can in principle be unambiguously assigned, but how confidently this can be done depends on how closely the senses are related and how much overlap their possible contexts share.

If we consider discreteness a property of perception rather than one of word meanings, the set of predetermined senses in a sense inventory can naturally be taken as a tool for construing meanings, analogous to using a ruler as an instrument for measuring length. One reads off a discrete number (e.g. 1.3 cm) from the ruler, but length itself is still a continuous function. Moreover, different measuring

---

[6] The "Bank Model" suggests that most words are like the word "bank" which has perfectly distinct senses. It predicts that each occurrence of "bank" would refer to one sense (a money bank) or the other (a river bank), which should be instant and effortless for English speakers to recognise. Kilgarriff (1993) tried to match each occurrence of his test words in the Lancaster Oslo-Bergen (LOB) corpus to one of their senses listed in LDOCE, and found that 87% (60 out of 69) of the test words had at least one usage which could not be confidently assigned only one sense, with which he concluded that word senses are by no means discrete. Wilks (1998) pointed out that the 87%-claim has no contradiction to saying that the majority of the text usage could be associated with only one dictionary sense.

instruments may be calibrated with different precisions (e.g. compare a ruler with a vernier caliper). Such discrete perception of continuous phenomena should not be unfamiliar to us. For example, the light spectrum is obviously non-discrete, but we, as laymen, identify red, yellow, blue, etc., as if they are fairly distinct colours. But for instance, blue covers a range of frequency and is sometimes more finely distinguished from light blue to deep blue. Some even discretise a kind of light blue as cyan, which corresponds to a narrower range of frequency than light blue in general on the light spectrum. We seldom disagree with one another between red and green, but perhaps it is more difficult to unanimously classify a somewhat bluish green as blue or green. Similarly, one might realise that when playing a string instrument, the same note may hardly be produced by pressing the string at precisely the same position every time. Slight difference within a certain range often goes unnoticed by the audience or even professional players.

Sense distinctions in individual sense inventories may be inexact, but as Wittgenstein (1958) suggested, "inexact" does not mean "unusable". Hence, the real limitation of grounding WSD on a set of predetermined senses does not lie in the discreteness arguments, but in the number and granularity of the senses listed in an inventory, which directly affect the predictability of the sense for a novel occurrence of a word. Although sense inventories vary on these dimensions, there should be some standard which one can expect. As Cruse (1986) suggested:

> "The number of fully established senses is presumably finite at any one time (though it may differ for different members of the language community, and at different times for the same speaker)." (p. 70)

1.5 Formulating the WSD Task

Now that we have clarified the discreteness assumption, we can formulate the task of word sense disambiguation as established in current mainstream practice. To start with, a *sense inventory* listing a finite set of predetermined senses for different words is needed. Such sense lists are often associated with some *lexical resources* which provide the lexico-semantic information of the senses, by means of definitions, semantic relations, or other forms of knowledge representation. Such lexical knowledge traditionally relies on the available lexical resources like machine readable dictionaries and semantic lexicons, but is more often obtained in sizeable amount from large corpora these days. We call such knowledge about the senses the *Conventionalised Context* (CC) of the senses. During disambiguation, should anything provide the hint for the sense of an ambiguous word, it must be the actual context embedding the occurrence of the word. We call this the *Triggering Context* (TC). Hence WSD can be considered the search for the best match between TC and CC from all the candidate senses.

Let us illustrate the process with Covington's (1994) textbook WSD algorithm as an example. To disambiguate the word "pen" in the sentence "The pen is full of pigs", we first have a lexical resource which provides the predetermined senses of "pen", including "a writing instrument" and "an enclosure for livestock", labelled as *pen_for_animals* and *pen_for_writing* respectively. In Covington's particular resource, senses are characterised by a group of cue words. Hence, the cue words for *pen_for_animals* include "farm", "pig", etc.; and those for *pen_for_writing* include "ink", "paper", etc. Since the CC in this case is in the form of cue words, we try to find matching cue words in the TC. Consequently, the match from "pigs" renders the sense *pen_for_animals* more likely.

Thus there are three critical components in WSD, namely lexical resource, contextual information, and the nature of target words. The lexical resource determines the number and granularity of the reference senses, which directly affect the difficulty of WSD. The best level of sense granularity, however, seems to be application-dependent, and recent research tends to suggest that most NLP applications need sense distinction up to the homonym level (e.g. Ide and Wilks 2006). The lexical information characterising the senses in the lexical resource also determines what kinds of disambiguating information are usable. Veronis (2001) commented that traditional dictionaries and other lexical resources alike often lack distributional criteria like syntactic and collocational information which are usually required to match a given sense with a new occurrence. Recent efforts have therefore resorted to obtaining disambiguating information from large corpora. These two extrinsic factors, that is, what makes a useful lexical resource and what kinds of contextual information are needed for WSD, have been extensively studied by computational linguists and will be reviewed in Chap. 2.

The nature of target words is a less straightforward issue. It happens that a word like "pen", with the two senses considered, can normally be distinguished via the presence of particular cue words. However, not all words have strongly associated cue words, and such words may require other types of information for effective disambiguation. According to Wittgenstein (1958), for example, words of different kinds are explained and understood by different mechanisms. Therefore, it is obvious that any one type of information will not work for all. As shown in recent WSD evaluation exercises, to be discussed in Chap. 3, multiple kinds of contextual information are indispensable in any WSD system, but the long realised *lexical sensitivity* of WSD still remains an open problem. We therefore aim at studying the behaviour of the target words so as to make effective use of various types of contextual information accordingly. To this end, we intend to revisit the cognitive aspects of WSD, which had once significantly informed the design of early WSD systems, for further insight on the lexical sensitivity issue. Chapters 4–6 will be devoted to this endeavour.

1.6 Overview of Subsequent Chapters

The current chapter has thus laid the background of the subject matter of this book by defining the problem of word sense disambiguation, discussing the nature of word senses, and explicating the task as established in mainstream practice.

In Chap. 2 (*Methods for Automatic WSD*), we will review the development and state of the art of WSD from the computational perspective, focusing on the knowledge sources contained in various forms of lexical resources employed in WSD, and the algorithmic approaches to exploiting such knowledge sources.

In Chap. 3 (*Lessons Learned from Evaluation*), we will discuss the history of WSD evaluation and summarise the lessons learned from the previous SENS-EVAL and SEMEVAL exercises. The positive and negative symptoms of WSD revealed from these evaluation exercises will be analysed.

In Chap. 4 (*The Psychology of WSD*), we will review the psycholinguistic evidence for how word sense ambiguities are handled by humans, including the representation of meanings in the mental lexicon and the factors affecting lexical access of words with multiple meanings.

In Chap. 5 (*Sense Concreteness and Lexical Activation*), we will explore the mental lexicon with respect to the concreteness and abstractness of concepts based on word association data, and investigate to what extent such concreteness distinction is modelled in existing lexical resources, and thus their potential roles in WSD.

In Chap. 6 (*Lexical Sensitivity of WSD: An Outlook*), we will discuss the potential use of concreteness as an alternative classification for word senses in terms of their information susceptibility in WSD. We will propose a lexically sensitive model of WSD and conclude with a research agenda to break new ground for automatic WSD research, to address the remaining unknowns by integrating more psycholinguistic evidence.

Chapter 2
Methods for Automatic WSD

Abstract Research in automatic word sense disambiguation has a long history on a par with computational linguistics itself. In this chapter, we take a two-dimensional approach to review the development and state of the art of the field, by the knowledge sources used for disambiguation on the one hand, and the algorithmic mechanisms with which the knowledge sources are actually deployed on the other. The trend for the latter is relatively clear, correlating closely with the historical development of many other natural language processing subtasks, where conventional knowledge-based methods gradually give way to scalable, corpus-based statistical and supervised methods. While the importance of multiple knowledge sources has been realised at the outset, their effective use in disambiguation systems has nevertheless been constrained by the notorious problem of "knowledge acquisition bottleneck" and is therefore very much dependent on the availability of suitable lexical resources.

2.1 A General Model

Given the long history of the field, the number of studies bearing on automatic word sense disambiguation (WSD) is enormous.[1] Reviews on WSD research often follow a more or less historical perspective to reveal the development from its early days to the present, which shows a very close correspondence between the

---

[1] The subject is also known by other names, e.g. word sense discrimination (McRoy 1992; Schütze 1998), lexical ambiguity resolution (Hirst 1987), automatic sense disambiguation (Lesk 1986), sense tagging (Wilks and Stevenson 1997), sense clustering (Chen and Chang 1998), word sense classification or supersense tagging (Ciaramita and Johnson 2003), word sense induction (Navigli and Crisafulli 2010), etc. Stevenson (2003) distinguishes sense disambiguation and sense tagging as different levels of WSD. However, as he also pointed out, the delimitation is not always clear-cut.

O. Y. Kwong, *New Perspectives on Computational and Cognitive Strategies for Word Sense Disambiguation*, SpringerBriefs in Speech Technology, DOI: 10.1007/978-1-4614-1320-2_2, © The Author(s) 2013

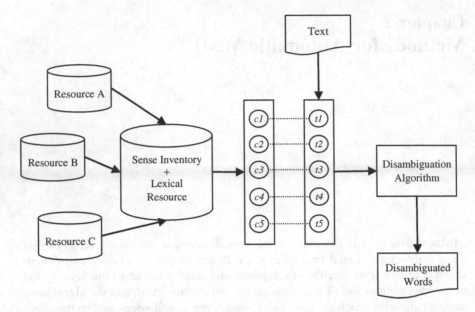

Fig. 2.1 A general model of WSD

kind of lexical resources available and the major algorithmic approaches adopted at any given period of time. Ide and Veronis (1998) did an exemplary survey of WSD development before the end of the last millennium, in which WSD methods are classified into AI-based, knowledge-based, and corpus-based methods with a common axis along the predominant kind of lexical resources during various historical periods. The individual chapters by Mihalcea (2006) on knowledge-based approaches, Màrquez et al. (2006) on supervised approaches and Pedersen (2006) on unsupervised approaches in the volume edited by Agirre and Edmonds (2006) all give a more recent and focused account of the major classes of WSD algorithms respectively. Navigli's (2009) comprehensive review in the *ACM Computing Surveys* gives detailed technical descriptions of particular algorithms within each class of methods applied in the context of WSD. Yarowsky's (2010) algorithm-oriented summary in the second edition of the *Handbook of Natural Language Processing* also serves as a succinct overview to the state of the art of WSD.

As mentioned in the last chapter, current practice of WSD is grounded on a sense repository to start with, and the task can be considered one which compares the *Triggering Context* (TC), that is, the actual context embedding a new occurrence of a word, with the *Conventionalised Context* (CC) characterising individual senses of the word, to find the sense with the closest resemblance between TC and CC. A general model of the task is shown in Fig. 2.1, and we will look at each component in the following sections.

2.1.1 Sense Inventory

Before the evaluation of WSD systems was standardised and converged to a few commonly used sense inventories (as will be discussed in Chap. 3), the following kinds of sense distinction have been adopted in different studies, roughly corresponding to their order of appearance in WSD history (except the last one):

- as in a customised sense inventory, which is closely associated with hand-crafted tailor-made lexical resources used in the early systems (e.g. Small and Rieger 1982; Hirst 1987; McRoy 1992);
- as translation difference in a second language, which used to be dichotomous sense distinctions for some specifically selected words in a source language with two words in a target language corresponding to its different senses (e.g. Brown et al. 1991; Gale et al. 1992a; Dagan and Itai 1994), and more often drawn from parallel corpora in recent studies (e.g. Diab and Resnik 2002; Ng et al. 2003);
- as simulated ambiguities like pseudo-words, which are artificially created examples of dichotomous sense distinctions solely for experimental purposes (e.g. Yarowsky 1993; Sanderson 1994; Dagan et al. 1997);
- as listed in existing lexical resources such as LDOCE (e.g. Cowie et al. 1992; Stevenson and Wilks 1999), WordNet (e.g. Leacock et al. 1993; Agirre and Rigau 1996; Ng and Lee 1996), and others, which is the most commonly adopted practice in contemporary systems, though it is often criticised for being too fine-grained to be realistically distinguished by WSD systems and of practical use for real language applications[2];
- as coarse-grained semantic categories such as the broad semantic classes in Roget's Thesaurus (e.g. Yarowsky 1992) or labels from the lexicographer files in WordNet, often called supersenses (e.g. Ciaramita and Johnson 2003; Curran 2005; Kohomban and Lee 2005; Yuret and Yatbaz 2010); and for the case of verb disambiguation, classes in Levin's (1993) verb classification have sometimes been used (e.g. Lapata and Brew 2004);
- as word clusters derived from texts, which are actually data-driven and thus dynamic sense distinctions, usually used with unsupervised learning methods treating WSD as a sense discrimination task and therefore not exactly as what is done in mainstream WSD (e.g. Lin 1998; Schütze 1998; Navigli and Crisafulli 2010).

---

[2] Thus in the past, some only considered a reduced set of senses (e.g. Leacock et al. 1993; Bruce and Wiebe 1994; Leacock et al. 1998; Towell and Voorhees 1998), or even only two distinct senses (e.g. Brown et al. 1991; Gale et al. 1992a; Yarowsky 1995), usually because of data availability and the respective methods used. The recent general understanding is that homonymy is more reasonably and usefully handled by WSD systems, and fine-grained senses in existing resources should better be merged to give more distinct groups of senses (e.g. Ide and Wilks 2006; McCarthy 2006).

2.1.2 Knowledge Sources

The disambiguation is subsequently based on a comparison between the TC containing a new occurrence of a target word against the CC characterising each potential sense of the word, usually by means of different kinds of linguistic information, or more often known as the *knowledge sources* for WSD. These are indicated by *c1...c5* and *t1...t5* in Fig. 2.1 for CC and TC respectively. Hence, *what* knowledge sources are actually compared is constrained by what is available as CC in the lexical resource(s) accessible by a system; whereas *how* they are compared depends on the algorithmic approaches adopted by individual systems. They thus form two essential dimensions in understanding WSD methods.

A wide range of knowledge sources has been employed in WSD,[3] constituting a spectrum of information across different linguistic levels, such as:

- *Pragmatic information*, as hand-coded in the form of contextual frames and inference rules in early systems (e.g. Wilks 1975a; Hirst 1987), or domain information included in some machine readable dictionaries like LDOCE as pragmatic/subject codes which is often used together with other types of knowledge (e.g. Stevenson and Wilks 2001).
- *Semantic information*, which involves the conceptual relations of different degrees of specificity among words, as one would expect to find among the different kinds of associations in a semantic network modelling the connections among concepts and words in the human mind.
- *Syntactic information*, including structural properties like syntactic relations from parses (e.g. Dagan and Itai 1994) or dependency relations (e.g. Lin 1997), subcategorisation frames (e.g. Martínez et al. 2002), and part-of-speech (POS) categories (e.g. Wilks and Stevenson 1996).

Given the nature of the task, semantic information often receives the most attention, while others are often considered weak knowledge sources and used in combination, in addition to semantic information (e.g. Stevenson 2003). We further categorise and focus on three major kinds of semantic information deemed important and central for WSD:

- *Broad semantic relations*, which are usually taken as conceptually related words in general and handled as a bag of words within a window embedding the target word. For example, one sense of "service" would be broadly related with words like "waiter" and "restaurant", while another would be related with words like "ball" and "player".

---

[3] In fact, some of the knowledge sources like selectional restrictions and subcategorisation frames may have a mutually beneficial relationship with WSD. On the one hand, these knowledge sources are useful for word sense disambiguation; and on the other hand, the ability to disambiguate word senses is also found to help the acquisition of selectional restriction and verb subcategorisation patterns (e.g. McCarthy 1997; Korhonen and Preiss 2003).

- *Narrow semantic relations*, which are some single, explicitly stated relationship between words, such as X IS-A Y, or A is PART-OF B, etc. For example, "service" is a kind of "tennis stroke". Such relations are therefore part of the broad semantic relations, but the precise relations and their representation allow more specific metrics for measuring the *similarity* between words, which is a more specialised form of *association*.
- *Positional semantic relations*, which are relatively local and hold between specific positions in a sentence, such as selectional restrictions, semantic roles and collocation patterns. For example, in "the dish is delicious", the adjective "delicious" tends to select for the "food item" sense rather than the "container" sense of the word "dish".

Another common term found in the WSD literature, especially on supervised learning methods, is *features*, which can be considered the algorithm- or system-specific realisation or representation of the knowledge sources used for disambiguation. The most common *features* used in WSD systems are categorised into *local features* which usually comprise n-grams of POS tags, lemmas or word forms with respect to the target word, *topical features* which correspond to the bag of words within a larger window, and *syntactic dependencies* which capture the syntactic cues and relations at the sentence level (e.g. Màrquez et al. 2006). Local and topical features thus encompass most of our three kinds of semantic relations, but differentiating them in terms of the distance from the target words instead of the nature of the disambiguating information. Alternatively, Specia et al. (2010) distinguished between shallow and deep knowledge sources. Shallow knowledge sources include topical word associations, collocations, etc., while phrasal verbs and selectional restrictions are considered deep knowledge sources. Individual knowledge sources could be realised in multiple ways. For example, topical word associations were obtained from overlapping words with dictionary definitions, bag of words from a context window, as well as frequent bigrams of adjacent words in a sentence.

In early studies, semantic information was often hand-coded for particular language understanding systems. Such lexical resources usually contain very rich information (often including world knowledge), but that also makes them time-consuming to form and thus severely limited in size. The information is also represented in very specific data structures and accessed with a pre-defined set of operations in these lexical resources, thus restricting their portability.

Wilks' (1975a, b) Preference Semantics System, for instance, represents lexical knowledge with constructs at three levels. "Semantic elements" consist of 70 primitive semantic units expressing entities, states, actions, etc. They are composed together with brackets and other syntactic constraints to give "semantic formulas", representing the senses of English words. "Semantic templates" define allowable patterns for combining semantic formulas into triples representing message patterns. The vocabulary size was about 500 and was already one of the largest of any operating deep-structure semantic analyser at that time.

In Small's Word Expert Parser (Small and Rieger 1982; Adriaens and Small 1988), each word is treated as a complex procedural knowledge source (i.e. "word expert"), in the form of a discrimination network with context probing questions at the nodes and the contextual meanings at the arcs. These word experts are defined by the Sense Discrimination Language and their interaction controlled by the Lexical Interaction Language. Each expert is of considerable size, like the one for "throw" was reported to be six pages long already but still expected to be ten times bigger, so the limited coverage of the lexicon can be imagined.

Both Hayes (1977) and Hirst (1987) represented knowledge in semantic networks. The former represented senses as nodes and semantic relations as links, and superimposed some frame-like structures (called "depictions") on them. The latter encoded the lexicon into a semantic network of frames, containing case slot restrictions and semantic relations like IS-A in each frame.

All of the above early work made serious attempts on the lexical resource itself for application in WSD. However, since hand-coding semantic information is heavy duty, the resultant resources were inevitably limited in scale. This problem is often known as the *knowledge acquisition bottleneck* for WSD. Section 2.2 will discuss how the availability of machine readable dictionaries and large corpora has enabled the acquisition of various knowledge sources in a much larger scale.

2.1.3 Disambiguation Mechanisms

Studies on human semantic processing have significantly inspired the design and implementation of WSD systems, which is particularly evident in the early WSD systems. Quillian (1968) proposed a computational model of human semantic memory. Apparently, the information stored in our memory is not isolated, but connected. So a model of semantic memory should have, in Quillian's words, "the ability to use information input in one frame of reference to answer questions in another". Thus his model has a network structure with a mass of nodes interconnected by associative links of different kinds, though in practice the actual substance in the network is as important as the formal structure of it.

Quillian's semantic network model allows two word concepts to be compared and contrasted via the association between them. This was demonstrated with a simulation program, which was able to say which meaning of each word was involved in each semantic similarity or contrast found—essentially what is expected of WSD. The association between concepts was found by expanding the search for related concepts from the two given concepts through the associative links and applying an activation tag to every concept reached, until an intersection (i.e. when two activation tags meet) was found. This method is generally known as "spreading activation with marker passing". Such network representation of semantic memory and the spreading activation model have generally been accepted as a plausible mechanism for human semantic processing given the support from experimental psychology on semantic priming (e.g. Collins and Loftus 1975), and have been implemented in many early WSD programs.

For example, in Hayes' (1977) system mentioned above, depictions (the frame-like structures) were connected by binders, and there were certain representational constructs to capture special cases of the subset-superset and part-whole relations. These structures functioned with corresponding association rules, so that only the relevant depictions and thus the appropriate senses for the words were activated.

Hirst (1987) also used marker passing with a semantic network of frames, but controlled the process with additional constraints on the number of propagations and on the nodes from which activation is to be spread. Parsing and disambiguation were synchronised with text input. In the event where no connection path was readily found for an ambiguous word to produce a definite analysis, Hirst introduced the use of Polaroid words. These were faked semantic objects with packets of knowledge used to substitute the actual analysis for a word temporarily until more information became available for eliminating the inappropriate senses.

Although these early systems demonstrated an obvious synergy between computational linguistics and psycholinguistics, only a minimal and diluted trace of the latter is usually found in state-of-the-art automatic WSD systems (further discussed in Sect. 2.3). We will, however, revisit the cognitive aspects of WSD in Chap. 4.

2.2 Knowledge Acquisition for WSD

Given the importance of multiple knowledge sources for WSD and the time and labour involved in crafting a sufficient lexical resource for the purpose, the availability of machine readable dictionaries and large corpora has brought hope to the knowledge acquisition bottleneck.

2.2.1 From Machine Readable Dictionaries

Despite the rich semantic information contained in dictionary definitions, they are nevertheless intended for human readers and are not readily usable by WSD systems. Relevant semantic information has to be extracted and represented in a computationally tractable way.

A widely researched aspect of acquiring semantic information from dictionary definitions is the construction of some taxonomy from these definitions. Amsler (1981), analysing the Merriam-Webster Pocket Dictionary, found that "each textual definition in the dictionary is syntactically a noun phrase or verb phrase with one or more kernel terms". Identifying these kernel terms makes it possible to assemble all headword-kernel pairs into a taxonomic semi-lattice, and Amsler's analysis resulted in lattices (linked by the IS-A relation) of 24,000 noun nodes and

11,000 verb nodes respectively. Subsequent studies have also used definitions from the Webster's Seventh Collegiate Dictionary (e.g. Chodorow et al. 1985; Markowitz et al. 1986; Klavans et al. 1990) or LDOCE (e.g. Nakamura and Nagao 1988; Vossen et al. 1989; Guthrie et al. 1990; Bruce and Guthrie 1992; Vossen and Copestake 1993). On the one hand, it is critical to be able to identify different defining formulas for accurate analysis and extensive coverage, as the general genus-differentiae definition pattern may have variations such as empty heads (e.g. Chodorow et al. 1985; Guthrie et al. 1990) or genus-headword relations other than simple IS-A (e.g. Markowitz et al. 1986; Nakamura and Nagao 1988; Vossen et al. 1989; Klavans et al. 1990). On the other hand, the genus terms need to be disambiguated and this was done by various means, such as by comparing the LDOCE box codes and subject codes between the genus terms and the headwords (e.g. Guthrie et al. 1990), or by deeper analysis of the syntactic and lexical patterns of the definitions (e.g. Klavans et al. 1990), etc.

Apart from narrow semantic relations, Veronis and Ide (1990) demonstrated the construction of very large neural networks (VLNNs) comprising broad semantic relations from the definitions in the Collins English Dictionary. A headword forms a node itself, and so does each of its senses. The word node is connected to its sense nodes by excitatory links, while each sense node is in turn connected by excitatory links to all word nodes of the words appearing in the definition of that sense. At the same time, senses of the same word are connected by inhibitory links. The connection process is repeated a number of times (for words in the definitions in the second round, and so forth). Veronis and Ide's final network consisted of a few thousand nodes with 10–20 thousand transitions.

2.2.2 Emergence of WordNet

In the early 1990s, WordNet (Miller et al. 1990; Fellbaum 1998) was made available to the natural language processing (NLP) community, on which it has exerted great impact. Developed at Princeton University, WordNet is probably the first broad coverage general computational lexical database. The project started as a psycholinguistic one for testing the scalability of relational lexical semantics, but the resulting database turned out to be of great interest to computational linguists. Words are grouped into concepts based on synonymy, forming "synsets" as the building blocks, which essentially represent the senses of words. In the latest version (3.1), there are approximately 117,000 synsets in the database. Synsets are linked to one another by relational pointers (e.g. hypernym, antonym, etc.), forming semantic nets, one for each of nouns, verbs, adjectives, and adverbs. For the noun database, synsets are primarily linked via hypernymy/hyponymy (i.e. IS-A relations) to form hierarchies with unique beginners which correspond to distinct semantic fields (e.g. "entity", "abstraction", "psychological feature", and others). Other relational pointers for noun synsets include holonym, meronym, antonym, and attribute. In earlier versions, paradigmatic associations between words are well captured through these pointers, but not syntagmatic relations. This deficiency has been remedied to a certain extent by the work on semantic tagging,[4] and on-going efforts are made to identify other relations among words. For instance, cross-POS relations including morphosemantic links (e.g. observe-observant-observation) and semantic role relations between verb and noun pairs (e.g. painter is the agent of paint) are available in recent versions.[5]

2.2.3 From Large Corpora

Broad semantic relations in the form of general co-occurrence or association can be estimated by simple conditional probabilities from large corpora. Church and Hanks (1990) identified certain significant syntactic and semantic co-occurrence patterns by means of mutual information I between two words x and y which have probabilities $P(x)$ and $P(y)$ respectively, which reflects the significance of the association between x and y:

$$I(x, y) \equiv \log_2 \frac{P(x, y)}{P(x)P(y)}$$

Generalising from such pairwise associations, co-occurring words of different topic categories can be learned especially from domain-specific corpora (e.g. Riloff and Shepherd 1999; Roark and Charniak 1998). Generally, the process starts with a few seed words for a given category (e.g. "vehicles", "weapons") and conditional probabilities of words co-occurring with them in particular syntactic structures (e.g. conjunctions, lists, etc.) with respect to the category are computed. Words at the top of the probability list will be chosen as seed words for the next round, and the process is repeated until enough words are collected for the category. Riloff and Jones (1999) showed that the noise ratio could partly be remedied by multi-level bootstrapping. Extending on this line, Caraballo (1999) attempted to develop a labelled, WordNet-like, hierarchy of nouns from co-occurring nouns within conjunctive and appositive structures in the Wall Street Journal (WSJ) corpus, though it was found to be less than straightforward with the high noise ratio and presence of sense ambiguity.

More recently, Ponzetto and Navigli (2010) automatically extended WordNet with large amounts of semantic relations from Wikipedia, making WordNet++. This was done by a mapping between Wikipedia pages and WordNet senses, followed by transferring the relations connecting Wikipedia pages to WordNet.

As an example for extracting positional semantic relations, Resnik (1993) learned selectional restrictions from parsed corpora by formalising selectional constraints as "an information-theoretic relationship between predicates and the

---

[4] See Landes et al. (1998) for the associated project on building semantic concordances.
[5] http://wordnet.princeton.edu/.

taxonomic classes of arguments", where the classes are modelled with the WordNet noun hierarchy. The fitness of a class of argument to a given predicate depends on the contrast between the posterior and prior distributions of the class. He thus defined the "selectional preference strength":

$$\eta_i = \sum_c p(c|p_i) \log \frac{p(c|p_i)}{p(c)}$$

and the "selectional association":

$$A(p_i, c) = \frac{1}{\eta_i} p(c|p_i) \log \frac{p(c|p_i)}{p(c)}$$

where p_i is a particular predicate and c is a class of argument. Ranking the selectional associations of a particular predicate with different classes of argument would reflect how strongly the predicate selects a class relative to others.

2.2.4 Combining Multiple Resources

While different types of lexical information might be acquired from individual resources, this may not be entirely satisfactory because a single resource may not contain or may not be good for all types of information. WSD typically requires a wide range of semantic information, so a better solution is to join various (types of) resources into an integrated repository. This is often treated as a task of sense mapping across various resources. For example, McHale and Crowter (1994) mapped each LDOCE sense to one of the 1,024 categories in the Roget's International Thesaurus by finding the best "measurement of relatedness" between the words in a sense definition in LDOCE and the potential categories of the target word in the thesaurus. Knight and Luk (1994) combined LDOCE and WordNet with a "Definition Match Algorithm". Kwong (1998) proposed a structurally based algorithm which indirectly links a dictionary and a thesaurus via a mediator resource. The sense mapping is first done from the dictionary to the mediator, and then from the mediator to the thesaurus. The algorithm was designed with types of resource rather than specific resources in mind, and was implemented for integrating LDOCE and Roget's Thesaurus, with WordNet as the mediator.

2.3 A Two-Dimensional Review of WSD Methods

As mentioned in Sect. 2.1.2, the lexical resources largely determine what knowledge sources are available for WSD, while the disambiguating algorithms control how they are actually deployed. In the following sections, these two dimensions will be considered simultaneously for reviewing a variety of methods in the WSD literature.

2.3.1 Knowledge-Based Methods

Machine readable dictionaries, semantic lexicons, and large electronic corpora have encoded particular lexico-semantic information in some way and the availability of these resources enabled the acquisition of different kinds of knowledge for characterising the senses of a word, as discussed in the last section. Knowledge-based WSD thus makes use of such lexical resources and the information therein to assign the best fitting sense to a new occurrence of a word.[6]

2.3.1.1 Broad Semantic Relations

When a word sense is expressed by a group of other words, word overlapping is a common way to measure the closeness between two word senses. Lesk (1986) was perhaps the first to use intact dictionary definitions for this purpose. WSD was done by finding the combination of senses (from all content words in a window) which had most overlaps of words in their definitions. Informal results of 50–70% accuracy were reported, using a default window of 10 words and definitions in the Oxford Advanced Learner's Dictionary of Current English.

Despite its simplicity and effectiveness, Lesk's method is deficient in several ways: (1) It is affected by the size of the dictionary and the exact wording of the definitions. (2) It is easily trapped by combinatorial explosion.[7] (3) Its "spreading activation" is limited to "first order", which may not be sufficient for uncovering more indirect relationships between words.

For the first problem, Wilks et al. (1989) and Guthrie et al. (1990) used co-occurrence data extracted from dictionary definitions to expand the definitions before counting overlaps, to increase the chance of finding some.

For the second problem, Cowie et al. (1992) applied simulated annealing to optimise the process. An energy function E is computed for a configuration C (i.e. a combination of senses). More overlaps of words in the definitions and subject codes in LDOCE result in a smaller E. The process starts with the combination of

---

[6] Some of the methods discussed in this section make use of knowledge acquired from large corpora and therefore involve statistical techniques. Since the probabilistic models are applied to the acquisition of particular knowledge sources like selectional preferences and subcategorisation patterns, we group them under knowledge-based methods and distinguish them from other corpus-based WSD methods which are really trained on corpus examples with sense information. The latter will be considered supervised methods as discussed in the next section.

[7] Wilks and Stevenson (1996) remarked that a 12-word sentence could give rise to more than 10^9 sense combinations to evaluate.

all first senses, and is repeated with new configurations a few thousand times, ending in the final stable configuration with minimised E for the disambiguated senses. Tests on 50 example sentences from LDOCE yielded 47% correct disambiguation at the sense level and 72% at the homograph level.

For the third problem, Veronis and Ide (1990) extended Lesk's method by building VLNNs from dictionary definitions, enabling more distant and thus even broader relationships to be captured. Traversing the network with the traditional spreading activation method for a few dozen cycles and with feedback and inhibition sent between nodes accordingly, the most strongly activated sense node for each word at the final stable state would suggest the disambiguated sense.

2.3.1.2 Narrow Semantic Relations

The simplest way for measuring the similarity between two nodes in a semantic hierarchy like the WordNet IS-A hierarchy is edge counting (Rada et al. 1989): the more intervening edges, the less similar the two concepts. However, given the peculiarity of the WordNet hierarchy (e.g. the links are not necessarily uniform), Resnik (1995a) found a correlation of only 0.6645 between simple edge counting and the similarity judgement by humans.[8]

Resnik (1995a) thus measured the similarity between concepts by "information content" based on the most informative concept in the WordNet hierarchy which subsumed both, and the probabilities of concepts were estimated from corpora. This method resulted in a correlation of 0.7911 with human judgement, which was better than simple edge counting. The measure was adapted to disambiguate related noun groupings (Resnik 1995b), where the most informative subsuming concept for each pair of words in the group was first found, and credits were assigned to the senses of the other words which were also subsumed by that concept. Evaluation on 125 words chosen from one Roget category gave an accuracy of about 60%.

Agirre and Rigau (1996) used "conceptual density" to quantify the distance between concepts. Based on their assumptions on the hierarchy, concepts in a deeper part or a denser part were ranked closer. The algorithm looked for the concept subsuming the target noun (in the middle of a window) which gave the highest conceptual density with the other nouns in the window. With window size of 30, tests on all nouns in four SEMCOR (sense-tagged Brown Corpus) files gave 43% precision and 34.2% recall at the sense level, and 53.9% precision and 42.8% recall at the WordNet lexicographer file level (comparable to homographs).[9]

---

[8] The human judgement scores were from Miller and Charles (1991).

[9] Precision and recall are common performance measures in NLP. See Sect. 3.2.3 for their definitions.

Mihalcea and Moldovan (1999a) extended the "conceptual density" measure to enable disambiguation of verb–noun (v–n) pairs. Conceptual density based on the words in common between the group of nouns appearing in the glosses of all verbs in the subhierarchy of a verb sense and the group of nouns in the subhierarchy of a noun sense in the pair was computed for each v–n sense combination. Testing on 200 pairs of v–n samples extracted from two SEMCOR files, accuracies of 86.5% for nouns and 67% for verbs were found.

2.3.1.3 Positional Semantic Relations

Wilks (1975a, b) was the earliest to exemplify the use of selectional restrictions for ambiguity resolution in the form of selectional preference. Semantic templates were used to match the heads of the semantic formulas into triples which linked up an agent, an action, and an object in a text fragment. Template matching alone might resolve some ambiguities. Other formulas in the fragment were then attached to the basic template if they satisfied the preferences expressed in the subparts of the formulas in the template, and the corresponding dependency was marked. The final network containing most dependencies would be the preferred interpretation of the fragment.

Some machine readable dictionaries (e.g. LDOCE, CIDE) represent selectional restrictions of verbs in the form of semantic codes, though they are usually inadequate to be used alone (see below for multiple knowledge sources).

Resnik (1997) combined statistical and knowledge-based methods for using selectional restrictions in WSD. Each candidate sense was assigned a score from the maximum selectional association between a predicate and the ancestors of the sense. The sense with the highest score was chosen. The 100 most strongly selecting predicates were identified from the untagged Brown Corpus for various syntactic relationships. Test instances containing these words were extracted from the tagged part of the corpus, and the noun arguments in these instances were disambiguated. Accuracies from 35.3% (for adjective–noun) to 44.3% (for verb–object) were reported, all outperforming the baseline of random choices. However, since many verbs and modifiers in English do not select strongly, selectional restrictions might better be used as a complement to other types of knowledge.

McCarthy and Carroll (2003) acquired both verb–noun and adjective–noun selectional preferences from SENSEVAL-2 datasets which were automatically pre-processed and parsed. Tree cut models were used to give estimates of the probabilistic distribution over the WordNet noun hierarchy, conditioned on a verb or adjective class and a grammatical relationship. Using the acquired selectional preferences together with the "one sense per discourse" heuristic[10] to address the problem of coverage, 51% precision, 23.2% recall and 45.5% coverage were obtained when testing on the SENSEVAL-2 English all-words task data.

---

[10] See Gale et al. (1992a) for the "one sense per discourse" property in WSD.

The results were reported to be similar to unsupervised systems but inferior to other systems using multiple knowledge sources.

2.3.1.4 Multiple Knowledge Sources

Early AI-based WSD systems had already employed multiple knowledge sources available from their internal lexical resources, despite their limited scales. For instance, each word expert in the sense discrimination network of the Word Expert Parser contained extensive knowledge of the word itself (Adriaens and Small 1988). However, we only became more conscious of the importance of multiple knowledge sources when they were distributed across various lexical resources.

Among the knowledge-based systems, McRoy (1992) was probably the first to seriously investigate this issue. She used several knowledge sources including syntactic tag, morphology, collocation, word association, and selectional preference for WSD within the TRUMP language understanding system. These different types of lexical information were hand-coded and distributed across a core lexicon (for general, coarse sense distinctions), a dynamic lexicon (for context-specific, finer sense distinctions), a concept hierarchy, a set of collocation patterns, and a set of cluster definitions. All core senses were first retrieved based on morphology. The obviously inappropriate core senses were then eliminated and specialised senses were triggered through a series of pre-processing steps, including tagging, collocation identification and cluster identification. Each piece of evidence was assigned a positive score for a match and a negative score otherwise. The sense with the highest sum from all evidence would be chosen.

Harley and Glennon (1997) also used additive weights to combine evidence from several knowledge sources, and reported 73% accuracy on tests with CIDE senses. Subsequent state-of-the-art supervised systems, as discussed below, mostly consider multiple knowledge sources in the form of various contextual features for disambiguation.

2.3.2 Supervised Learning Methods

In *supervised learning*, ideally a large training corpus is available with every instance of a target word tagged with the appropriate sense. Knowledge sources at various abstract linguistic levels for disambiguation are rendered in the form of contextual features. As mentioned in Sect. 2.1.2, such features can be topical, local, or syntactic, depending on the window embedding the target word from which they are extracted. Machine learning algorithms are applied to train classifiers with contextual features extracted from sense-tagged examples, taking the individual senses as classes, and during disambiguation, the classifier will assign a new occurrence of a word to one of its senses based on the same feature set extracted from its triggering context.

2.3.2.1 Broad Semantic Relations

Broad semantic relations are often modelled by topical and local features capturing simple co-occurrence in the training data, and parameters are estimated from all words w co-occurring with the target sense s within a given context size. A Bayesian classifier typically disambiguates word senses by finding

$$\underset{s \in S}{\text{argmax}} \prod_{w\,in\,context} \frac{p(w|s)p(s)}{p(w)}$$

where S is the set of all candidate senses of a target word. Gale et al. (1992a) tested this method on six polysemous nouns (two senses each in terms of French translations), using a 100-word context, and obtained accuracies from 82% to 100%. They also observed that multiple occurrences of the same word in the same discourse always had the same sense, and this "one sense per discourse" property was exploited in later studies, e.g. Yarowsky (1995). Magnini et al. (2002) further proposes "one domain per discourse" and showed the importance of domain information for WSD.

Supervised learning requires large amount of sense-tagged data to be reliable. Leacock et al. (1993) compared three supervised learning methods, Bayesian classifier, content vector classifier, and neural network, for disambiguating the noun "line", with examples from WSJ for 6 out of 25 of its senses in WordNet. They found that the performance for all methods improved with the number of training examples. All reached about 70% accuracy at 200 training samples. However, hand tagging is not only labour intensive but also error-prone.

Alternatives to hand-tagging are thus sought. For example, Brown et al. (1991) and Gale et al. (1992a) replaced hand-tagging with the different translations in a second language from bilingual corpora. Class-based sense distinction is another way. Yarowsky (1992) took all instances for words in the same Roget category as training examples for a same sense. Using the Bayesian method and ±50 words as context to disambiguate 12 polysemous nouns found in the WSD literature yielded an average accuracy of 92%. Mihalcea and Moldovan (1999b) suggested that sense-tagged corpora be acquired by searching the Internet, using queries with synonyms and definitions of the various senses of a word, and replacing the search phrases with the original word (sense-labelled correspondingly) in the retrieved examples. Manually checking 1,080 examples (for 120 different senses) retrieved from the top ranked documents gave 91% accuracy.

2.3.2.2 Narrow Semantic Relations

Paradigmatic associations are often bound to specific lexical resources like WordNet, and they are most explicitly seen in knowledge-based WSD systems, like the ones on edge counting and conceptual density as discussed above. Since such narrow semantic relations are not directly available from tagged corpora, they are almost never seen in supervised learning methods. The analysis by Agirre and Stevenson (2006) also supports this observation. Rather, local and topical contextual features based on words are simply assumed to contain this kind of information amongst other broader relations, and are often used in combination with other features in supervised systems.

2.3.2.3 Positional Semantic Relations

Local collocation is generally defined as the use of words in constant patterns, which are not necessarily idioms but more like habitual usages. Unlike selectional restrictions, local collocations may or may not require explicit syntactic constraints. Yarowsky (1993) treated collocation as "co-occurrence of two words in some defined relationship" where the relationships were defined by positions. He tested several types of collocation like "content word to immediate right" and "first content word to left". Binary ambiguities in various forms (e.g. multiple translations, pseudo-words, homophones, etc.) and for different POS (nouns, verbs, and adjectives) were tested. Mean entropies of $P_r(Sense \mid Collocation)$ were computed from the training samples for every sense pair and each observed collocation. If a specific word in a given collocation for a test sample was among what was learned, the most probable sense was assigned, otherwise the most frequent sense was returned. An average precision of 95% (though at partial recall of about 50%) was found, and it was concluded that the same sense tended to hold for more than 90% of the time with a given specific collocation—the "one sense per collocation" property. It was observed that local collocation could capture some infrequent but useful evidence which was otherwise unavailable, and different POS may favour different collocation types and widths of context.

2.3.2.4 Multiple Knowledge Sources

Almost without exception, state-of-the-art supervised WSD systems, regardless of their actual learning algorithm, make use of multiple knowledge sources realised through a variety of contextual features.[11] This is most evident from the top performing systems in SENSEVAL.

An important consideration with the use of multiple knowledge sources is how they should be combined and how to handle conflicting preferences. The most straightforward way is to additively weigh individual pieces of evidence. For example, Leacock et al. (1998) built a Bayesian topical/local classifier which combines topical cues from open-class words found in a wide window and local cues including open-class words, closed-class items, and POS tags found in a

---

[11] Slightly different is Pedersen (2000), who used one type of information, namely co-occurring words, but combined evidence from various window sizes.

narrow window around the target word. Tests on examples of the verb "serve", the adjective "hard", and the noun "line" yielded 83–84% accuracy using topical and local cues together, which was better than either topical or local cues alone (except for the adjective "hard", where the combined effect was similar to that of local cues alone).

Lapata and Brew (2004) first acquired subcategorisation frames compatible with Levin's (1993) verb classification from the British National Corpus, and used these frame preferences as useful priors with other collocation and co-occurrence features in a Naïve Bayes classifier for disambiguating verbs with respect to the Levin verb classes.

While the Bayesian classifiers treated the various knowledge sources as independent of one another, Bruce and Wiebe (1994) attempted to systematically identify the interdependencies among them from decomposable models, retaining only the most important interdependencies among the features. Initial tests on the word "interest" (for six of its senses in LDOCE) achieved about 78% accuracy. Although comparable results were obtained for several other words (including other POS), the method seemed to work best with nouns. The biggest limitation, however, is the need for large amount of sense-tagged samples. Also, the feature values need to be simple, often dichotomous, and as the number of features increases, the method may become impractical when there are too many models to choose from.

Ng and Lee (1996) used the k-Nearest-Neighbour approach, which is an exemplar-based method, to compare a test sample to each of the training samples with respect to a feature set for the closest match, determined by the sum of the distances between the values of all the features of two samples. They reported 54% accuracy for testing on the Brown Corpus with WordNet senses. Stevenson (2003) used a similar approach, or memory-based learning, to combine multiple weak knowledge sources for sense tagging. His system contained several modules, including a POS filter, three partial taggers, and a feature extractor. These modules were combined with the TiMBL memory-based learning system[12] trained on sense-tagged examples. The algorithm attempts to classify a new instance by comparing it with the seen examples to find the closest ones and labeling it with the class to which these examples belong.[13] About 90% accuracy was reported for testing on all content words in SEMCOR with WordNet senses mapped to LDOCE senses. Mihalcea's (2002) is another example-based system, which is also a typical hybrid system containing two modules. One module uses patterns learned from

---

[12] http://ilk.uvt.nl/timbl/.

[13] Actually Stevenson's (2003) system is more appropriately considered a hybrid system since most of the individual modules have a very prominent knowledge-based element while the supervised learning part serves to conveniently combine them. For example, the three partial taggers work on word overlaps with dictionary definitions using simulated annealing, broad context based on subject areas of words indicated by the pragmatic codes in LDOCE, and selectional restriction information found in LDOCE as expressed by semantic codes in the dictionary, respectively.

existing resources including WordNet, SEMCOR, and sense-tagged corpora to tag all words in open text; and the other uses an instance-based learning algorithm with automatic feature selection to tag words with large amount of training data. The system achieved about 70% accuracy in the SENSEVAL-2 English all-words task and lexical sample task.

Decision lists provide another way to get an optimal combination of multiple knowledge sources. For example, Yarowsky (1994) trained a decision list with local syntactic patterns and more distant collocation evidence for restoring the accent patterns in Spanish and French words, which is treated as a special case of WSD. Collocational distributions (for various collocational patterns) were first measured, and the decision list was obtained by computing and sorting the log-likelihoods as follows:

$$Abs\left(Log\left(\frac{\Pr(Accent\_Pattern_1 | Collocation_i)}{\Pr(Accent\_Pattern_2 | Collocation_i)} \right) \right)$$

It was suggested that decision lists identify and utilise the single best disambiguating evidence in a target context, and therefore avoid the complex modelling of statistical dependencies amongst the knowledge sources. Wilks and Stevenson (1998), on the other hand, marked each sense of a target word as either appropriate or inappropriate in the context of each training example containing a feature vector of several knowledge sources. A decision list was learned from these examples and applied to disambiguate words in novel contexts. General rules were also produced alongside the decision list so that target words not covered by the training data could also be handled.

Wu et al. (2004) showed that a system based on Kernel Principal Component Analysis (KPCA), using position-sensitive, syntactic, and local collocational features, outperformed Naïve Bayes model, maximum entropy model, and Support Vector Machines. KPCA achieved 65.8% accuracy and the others 63.3%, 63.8% and 65.2% respectively, testing on the SENSEVAL-2 English lexical sample task dataset. The advantage of kernel methods is probably their "implicit consideration of combinations of feature information in the data distribution from high-dimensional training vectors". Gliozzo et al. (2005) used kernel functions for supervised WSD for modelling domain and syntagmatic aspects of sense distinctions, with the domain models acquired from external untagged corpora.

Ciaramita and Johnson (2003) trained a multiclass perceptron classifier using collocation, spelling and syntactic contextual features for classifying word senses of unknown nouns with respect to the WordNet lexicographer file labels (that is, supersense tagging). The training examples were extracted from the BLLIP corpus which is a 40-million-word syntactically parsed corpus based on the Wall Street Journal corpus, available from the Linguistic Data Consortium.

Dang and Palmer (2005) focused on the importance of semantic role information, using it with other topical and local features with a maximum entropy model for disambiguating the verb instances in the SENSEVAL-2 lexical sample task dataset.

Specia et al. (2010) used inductive logic programming as the supervised algorithm, taking advantage of its ability to induce hypotheses from examples and synthesise new knowledge from experience as a machine learning tool, and its representation formalism based on first-order logic for the rendition of more complex and deep knowledge sources beyond simple features. Their system used nine shallow and deep knowledge sources to disambiguate verbs and nouns. The knowledge sources were realised from information available in dictionaries, syntactic parses, and corpus data, and represented in the form of predicates. Models were created from the sense-tagged examples and background knowledge through several iterations, resulting in an ordered list of symbolic rules to be applied in sequence during disambiguation.

Lee and Ng (2002) compared the performance of several machine learning algorithms including Support Vector Machines (SVM), Naïve Bayes, AdaBoost, and decision trees.[14] They tested the systems on SENSEVAL-1 and SENSEVAL-2 data, using POS, words, and collocations in the local context as well as syntactic relations for the features. The best result was obtained from SVM with a combination of all knowledge sources, which was even better than the best official scores in SENSEVAL-1 and SENSEVAL-2. Nevertheless, it was also observed that different knowledge sources seem to work best with different learning algorithms. As reported, for example, local collocations and POS contribute the most for SVM and Naïve Bayes respectively.

Yarowsky and Florian (2002) did a comprehensive study on relative system performance across different training and data conditions with SENSEVAL-2 data on four languages. Similarly, interaction between feature sets, training sizes, and learning algorithms was observed. For example, aggregative algorithms like Naïve Bayes tend to perform well with bag-of-word features, while discriminative algorithms like decision trees tend to perform well with local collocations or syntactic features. Some algorithms are more tolerant than others of sparse data, high degree of polysemy and noise in the training data.

Màrquez et al. (2006) compared five significant machine learning algorithms used in previous studies: Naïve Bayes (NB), k-Nearest-Neighbour (kNN), Decision Lists (DL), AdaBoost (AB), and Support Vector Machines (SVM), trained on the same set of data. A group of 21 words was selected from the DSO corpus for this experiment. These selected words (13 nouns and 8 verbs) are commonly seen in WSD literature. The number of examples varies among words, and the average is 801.1 examples per word. Average number of senses per word is 10.1. Disambiguating information used includes 15 local feature patterns (with words or POS) and topical context in the form of bag of words (content words in the sentence). The most-frequent-sense classifier (MFC) was used as baseline. The following results were found:

---

[14] SVM attempts to find a hyperplane with the largest margin that separates training examples into two classes. AdaBoost attempts to boost the performance of an ensemble of weak learners by giving more weights to misclassified training examples so that the classifier will concentrate on these hard examples.

$$SVM \approx AB > kNN > NB \approx DL > MFC \,(p.192)$$

The accuracy ranged from 67.07% for SVM to 61.34% for DL, all of which outperformed the baseline of 46.55% for MFC. SVM and AB performed significantly better than kNN, which in turn performed significantly better than NB and DL. More interestingly, Màrquez et al. also compared the various methods in terms of the agreement on their predictions with the Kappa statistic, a measure commonly used for inter-annotator agreement with the value 1 indicating perfect agreement. It was found that SVM and AB were the methods that best learned the behaviour of the DSO examples. Also, the three highest values of Kappa were between the top performing methods (SVM, AB and kNN), and although NB and DL achieved a very similar accuracy on the corpus, their predictions were quite different as the Kappa value between them was the lowest.

The recent trend is thus not only to use multiple knowledge sources but also to use multiple classifiers in combination (e.g. Florian et al. 2002). For instance, as pointed out by Mihalcea et al. (2004), several of the top performing systems among the 47 participating systems in the SENSEVAL-3 English lexical sample task combine multiple classifiers with voting schemes, and such combinations often outperform individual classifiers. These developments will be further discussed in Chap. 3.

2.3.3 Unsupervised Learning Methods

The knowledge acquisition bottleneck in automatic WSD has stayed throughout the knowledge-based era to the supervised learning age. The main difficulty for the latter is that supervised algorithms often need a large amount of sense-tagged examples for training, but hand-tagging large corpora with sense information is time-consuming and error-prone. For example, Yarowsky (1995) took three years to hand-tag 37,232 examples, and Ng and Lee (1996) reported only 57% for inter-tagger agreement. Ng (1997) even projected an appalling figure of 16 man-years for sense-tagging a 3.2M corpus[15] which would be needed for any broad coverage, high accuracy WSD program. This leads to proposals on cheaper ways to obtain larger amounts of sense-tagged data, as well as unsupervised learning from untagged corpora. However, since unsupervised methods differ from mainstream practice and often treat WSD as a clustering or discrimination task, we will only briefly mention some examples in this section.

Yarowsky (1995) started with a small number of seed collocations (that is, representative examples) for each sense of an ambiguous word and trained

---

[15] The estimation was based on 3,200 most frequent words in the Brown Corpus which cover 90% of all word occurrences, and each should have 1,000 instances tagged.

decision lists for disambiguation.[16] More examples were incrementally acquired from the residual set until the residual set became stable. Testing 12 words (two senses each from French translations) gave an average accuracy of about 95%, which was comparable to that achieved with supervised methods, but the sense distinction was only dichotomous. Using a similar approach with neural networks as the learning algorithm, Towell and Voorhees (1998) found considerable distance between their results and the upper bound (i.e. with all examples manually labelled). Leacock et al. (1998) did unsupervised training from examples of the monosemous relatives of the target words as defined in WordNet. Four of the six test words attained similar results as training from manually tagged examples, but the other two got substantially lower results. It was suggested that some monosemous relatives might be too specific.

Karov and Edelman (1998) avoided manual tagging altogether by augmenting the training set (i.e. sentences of a word W to be disambiguated) with a feedback set (i.e. sentences of the nouns found in the entry of different senses of W in some machine readable dictionary). Each original sentence in the training set was finally tagged with the same sense as its most similar sentence in the feedback set, after an iterative learning phase. The similarity measure was based on the circular definition that words are similar if they appear in similar contexts and contexts are similar if they contain similar words. Sentences attracted to the same feedback set were clustered to give the typical use of that sense, and new examples were disambiguated by finding the most similar cluster, i.e. a similarity-based method. Test instances were from the Treebank-2 WSJ corpus; and the online versions of the Webster's and Oxford dictionaries as well as WordNet were used in combination. Tests on four polysemous words (two senses each, altogether 500 samples) gave 92% accuracy. Lin (1997) also used a similarity-based method to cluster word senses. Examples were not sense-tagged but were parsed with dependency information. Contextual similarity was thus measured by comparing dependency triples which contained two words and a dependency relation.

Pedersen and Bruce (1998) used the Expectation Maximisation algorithm to estimate the parameters for a Naïve Bayes model from a raw corpus for WSD. The features used include morphology, POS of neighbouring words, frequent co-occurring content words, unrestricted collocations, and the content word on the immediate left and right of the target word.

In Schütze's (1998) Word Space Model, all words co-occurring with a word w in a fixed-sized context have an entry in the word vector of w, which indicates the number of times the word co-occurs with w in the corpus. Within such "Word Space" containing multi-dimensional vectors, the semantic relatedness between two words can be crudely measured by a cosine function, and a context vector is given by the centroid of the word vectors occurring in the context. The component

---

[16] Methods like this which use a few examples for bootstrapping are sometimes counted as weakly/lightly/minimally supervised approaches, in contrast to genuine unsupervised methods which take no tagged training examples at all.

of a context vector with respect to a certain dimension thus reflects the topical strength of the context for that dimension. The context vectors of an ambiguous word are clustered into a predetermined number of groups, and the centroids of these context groups thus represent the senses of the word. WSD is handled as the discrimination among context groups, that is, to find the sense vector of the word which is closest to the context vector of its new occurrence.

Diab and Resnik (2002), on the other hand, exploited the translation correspondences in parallel corpora for unsupervised WSD. They first aligned parallel corpora at the word level, and then for a source word (e.g. "catastrophe" in French), identified a target set with all sentences containing an alignment with it (e.g. English sentences with "disaster" or "tragedy"). The target set was thus treated as monolingual WSD with respect to the target language sense inventory, and the sense tag of the target set was finally projected to the source word. The method was evaluated with pseudo-translation created by machine translation systems and results comparable to or even better than other supervised systems in SENSEVAL-2 all-words task were reported.

Yuret and Yatbaz (2010) applied a probabilistic generative model, namely the noisy channel model, for unsupervised WSD. Context-dependent sense distributions based on semantic classes were estimated from unlabeled text data. Each context C is modeled as a distinct channel where the intended message is a word sense (a semantic class) S, and the signal received is an ambiguous word W. Thus WSD is to find the sense S which maximises $P(S|W,C)$, equivalent to maximising $P(W|S,C)P(S|C)$. Since there is no sense-tagged data, word frequency data for each sense $P(W|S)$ and word frequency for the given context $P(W|C)$ are used for estimating $P(S|C)$ indirectly. The method was tested on all noun instances in the all-words tasks in recent SENSEVAL exercises, disambiguating them with respect to the 25 WordNet lexicographer file labels as supersenses. It was found that frequent concrete classes like person and body were disambiguated well, while the abstract classes were responsible for most of the errors. Supersense tagging, especially when done in an unsupervised setting, is often taken as a way for thesaurus construction to expand existing semantic lexicons like WordNet (e.g. Ciaramita and Johnson 2003; Curran 2005).

2.4 Summary

The success of contemporary WSD systems thus depends on the kind of knowledge sources used for disambiguation and the way they are actually deployed. Knowledge-based algorithms rely on the comprehensiveness of knowledge sources acquired from different lexical resources. Supervised learning methods often face the data sparseness problem and classifiers may have different bias toward the data. Understanding the individual and combined effects of knowledge sources as well as algorithmic approaches is essential, which will be the focus of our next chapter.

Chapter 3
Lessons Learned from Evaluation

Abstract The performance evaluation of word sense disambiguation systems has only been more or less standardised in the last decade with the first three SENSEVAL and the more recent SEMEVAL exercises. These exercises have pointed to the superiority of supervised methods using multiple knowledge sources and ensembles of classifiers. Behind the apparently plateaued performance of state-of-the-art systems, some fundamental issues including sense granularity, sparseness of sense-tagged data, and contribution to real applications, still remain. But more importantly, evaluation results also suggest that there is something about the target words themselves which is responsible for the differential performance among systems trained on the same feature sets, and within systems on different target words. We suggest that such intrinsic nature of the target words has a direct impact on the accuracy of sense disambiguation, which cannot be addressed solely from the computational perspective.

3.1 Post-Millennium Development

Word sense disambiguation (WSD) systems, as reviewed in the last chapter, can be knowledge-based, supervised, or unsupervised. These major approaches are also seen in the development of other areas of natural language processing (NLP) in general. The algorithms used in supervised and unsupervised learning are mainly generic machine learning methods applied in the context of WSD, and most of them have already been studied by the 1990s.

Ide and Veronis (1998) concluded in their survey in the introduction of the special issue of *Computational Linguistics* on WSD with three open problems which should direct future development of WSD, especially after half a century of research on the topic:

O. Y. Kwong, *New Perspectives on Computational and Cognitive Strategies*
for Word Sense Disambiguation, SpringerBriefs in Speech Technology,
DOI: 10.1007/978-1-4614-1320-2_3, © The Author(s) 2013

- *The Role of context*—They pointed out that WSD systems rely on the context of the target word to provide the disambiguating information, and context is often used in a bag-of-words manner or considered in terms of some relation to the target word. They further mentioned that information from microcontext (that is, local context as we have discussed), topical context, and domain contributes to WSD, but systematic study is needed to better understand the relative roles and importance of the various kinds of information, and their interrelations.
- *Sense Division*—Despite the imperfection of the "Bank Model", most WSD researchers have relied on the sense distinctions in existing lexical resources, typically machine readable dictionaries or WordNet. However, the appropriate degree of sense granularity remains one of the foremost problems, as many have found that the sense divisions in dictionaries are often too fine-grained for NLP and they might just be unrealistic for automatic WSD.
- *Evaluation*—It has been difficult to compare WSD results from previous studies because they often refer to different sense inventories with varied sense granularity and use different test materials. It is therefore important to be able to evaluate systems on a common platform, with reference to a manually sense-tagged benchmarking corpus and a set of sense distinctions agreed by the research community at large. In addition to in vitro evaluation, it is also necessary to see how WSD, as an "intermediate task", can improve the results for real language applications.

Some serious efforts have been made by researchers in the last decade or so to address these open problems of the field. The work on evaluation, to be introduced in Sect. 3.2, has matured most noticeably and gained considerable achievement, which is evident from the standardisation of assessing WSD systems with common datasets and well-defined tasks and procedures in the series of SENSEVAL exercises. The sense division problem has attracted more and more attention. Although we should not expect any one-off and discrete solution, especially given the many linguistic and philosophical issues involved in sense distinctions, the general consensus to date is that homograph is probably the optimal granularity which can be realistically handled by WSD systems. We will look at some ways by which coarser-grained senses are obtained in Sect. 3.3. As for the role of context, in addition to what is learned from SENSEVAL results, some studies have done comprehensive evaluation of the contributions of various knowledge sources and the effectiveness of different machine learning algorithms. However, while such studies reveal the complexity of the issue, the contribution of different types of disambiguating information to different types of target words has not been sufficiently addressed. We will explain what we think is missing in Sect. 3.4.

3.2 System Evaluation

As shown in Chap. 2, the experimental conditions of individual early WSD studies used to vary in so many different ways that comparison between systems is often difficult and fruitless. For instance, the sense inventories may differ in the senses they list for a word and the granularity by which they distinguish these senses. Some studies follow the senses listed in machine readable dictionaries (e.g. Stevenson and Wilks 2001), while others focus on dichotomous and very distinct senses of selected words (e.g. Yarowsky 1992), and the latter often report apparently better accuracy. Systems are evaluated on a variety of test materials. Some focus on examples for a handful of selected ambiguous words (e.g. Leacock et al. 1998), while others attempt disambiguation for all content words in a running text (e.g. Ng and Lee 1996), and the latter is often deemed more difficult. Some standardisation on these two factors is therefore necessary to provide a common platform on which the precise effects of various combinations of knowledge sources and disambiguation algorithms used in different systems can be fairly evaluated and compared.

3.2.1 Sense Distinctions

On the one hand, the number and granularity of senses in a particular sense inventory have a direct impact on the difficulty of disambiguation. On the other hand, such difficulty level should be taken into account when evaluating system performance for a fair comparison. Evaluating and comparing individual systems entirely by the number of correct predictions they make may have overlooked the more subtle differences with respect to the seriousness of their errors. For instance, despite its wrong predictions, a system should be considered better than others if it manages to successfully rule out some obviously wrong senses in most cases instead of spreading the errors across all other senses. Similarly, a system should be less penalised for its wrong predictions if they are semantically close to the expected answers than if they are very different senses. As Resnik and Yarowsky (1997) suggested, a simple metric to realise such distance-based scoring can be like this:

$$\frac{1}{N} \sum_{i=1}^{N} distance(cs_i, as_i)$$

and it can be used to minimise the mean distance between the assigned sense (as_i) and correct sense (cs_i) over all N examples, where the pairwise distance (for the senses organised in a sense hierarchy) could be derived from psycholinguistic data on similarity or confusability amongst the many possible sources.

3.2.2 Test Materials

Annotated corpora and artificial corpora have been used for evaluating WSD systems in the past (e.g. Stevenson 2003; Palmer et al. 2006). Typical sense-annotated corpora include the "interest" corpus (e.g. Bruce and Wiebe 1994), the

Table 3.1 Typical sense-tagged corpora for WSD

Data	Content
"Interest" corpus	2,369 instances of the word "interest" from the Wall Street Journal section of the Penn TreeBank, marked with one of its six nominal senses from LDOCE.
"Line-hard-serve" corpus	About 4,000 instances each of the noun "line", the verb "serve", and the adjective "hard", tagged with one of their six, four, and three possible senses respectively from a subset of their WordNet senses. The examples for "line" and "serve" were extracted from the Wall Street Journal and the American Printing House for the Blind corpus, and those for "hard" were taken from the San Jose Mercury News corpus.
DSO corpus	192,800 instances of 121 most frequent nouns and 70 most frequent verbs from the Brown Corpus and the Wall Street Journal corpus, tagged with senses from WordNet. Up to 1,500 examples for each word type are available.
SEMCOR corpus	Semantic concordances based on the texts in the Brown Corpus created alongside the WordNet project. All content words in 186 text files (about 2,000 words each) and all verbs in another 166 files were lemmatised and tagged with WordNet senses. The files cover different types of text, e.g. fiction, press reportage, scientific writings, etc.

"line-hard-serve" corpus (e.g. Leacock et al. 1998; Towell and Voorhees 1998), the DSO corpus (e.g. Ng and Lee 1996), and the SEMCOR corpus (e.g. Landes et al. 1998; Stevenson 2003), as described in Table 3.1. Artificial test data include those handling sense ambiguity in the form of translation difference (e.g. Gale et al. 1992a) or pseudo-words (e.g. Yarowsky 1993).

3.2.3 SENSEVAL and SEMEVAL

Assessment of WSD systems can be qualitative or quantitative. Early studies, testing only a few selected samples, did not usually opt for serious quantitative evaluation (e.g. Wilks 1975b; Hirst 1987; Adriaens and Small 1988; McRoy 1992). Studies from the 1990s onward tend to test much larger datasets and present results numerically (e.g. Agirre and Rigau 1996; Ng and Lee 1996; Harley and Glennon 1997; Stevenson and Wilks 1999). Numerical results are commonly reported by a simple accuracy figure (e.g. Lesk 1986; Veronis and Ide 1990; Yarowsky 1995), or in terms of precision and recall (e.g. Agirre and Rigau 1996; Rigau et al. 1997; Leacock et al. 1998). It may be more convenient to compare numbers, but sometimes even individual numbers may not be directly comparable. Also, qualitative analysis should not be less important, especially if the task is "highly lexically sensitive" as Resnik and Yarowsky (1997) suggested.

The lack of consensus about evaluation is a notorious problem for inter-study comparison. In fact, given the many necessary considerations for a fair comparison (e.g. whether to force every system to use the same resources, should the evaluation be task-dependent, how coarse/fine a kind of WSD is needed, whether to assume WSD as a one-place process in a larger system, etc.), it is not easy to reach a consensus. Gale et al. (1992b) proposed a relative scale for measuring WSD performance by estimating a lower bound and an upper bound. The lower bound was established by assigning each word its most likely sense. This, for instance, ranged from 48% to 96% for the dozen test words in Yarowsky (1992). The upper bound was derived from agreement among human judges. In view of the difficulty of a classification task like Jorgensen's (1990) which only resulted in 68% agreement, a discrimination task was used instead. Judges were asked to say whether the target words, with their definitions provided, in each pair of concordance lines (altogether 82 pairs of 9 polysemous words) had the same sense. On average 96.8% agreement was obtained among the five judges. However, Wilks (1998) said that methodologically it was pointless to include the definitions because the judges did not need them to do the task, and the results only showed whether the judges were sense "lumpers" or "splitters". We also believe that a valid upper bound can be estimated only when the judges share a similar task as a WSD system. In that case a recognition task, in which judges actually need to say which definition applies, would be more suitable.

Given proper estimations, the upper and lower bounds offer a relative scale for systems to compare with one another indirectly. Alternatively, to allow more direct comparison, Yarowsky (1992) selected test words which were used in other studies. Agirre and Rigau (1996), with a similar intention, additionally implemented some methods from other studies and tested those methods on their own dataset. However, this kind of "translation" between data or methods rarely preserves all the inherent characteristics of individual systems. Most notably, the set of candidate senses is not always the same across studies. Therefore, a framework for common evaluation and test set generation is necessary, as saliently remarked in Resnik and Yarowsky (1997). In fact, in addition to the need for a common evaluation platform, they also made recommendations to scoring methods. For example, their suggestions include a cross-entropy kind of evaluation criterion to take the eliminative ability of the systems into account, penalty adjustment from the confusability of sub-senses, and tackling the long-standing sense inventory problem by defining senses based on multi-lingual data. The last suggestion was further explored in Resnik and Yarowsky (1999). Their proposal catalysed the inception of SENSEVAL[1] for the first time in 1998, where their suggestions regarding the size and sampling of the dataset, the task difficulty for the annotators, and the accommodation of both supervised and unsupervised systems, were realised (Kilgarriff and Palmer 2000).

Strongly motivated by the need to standardise the test set, a most remarkable achievement of the SENSEVAL project for WSD evaluation is thus the production of a "gold standard" corpus tagged with correct answers as the shared dataset (Kilgarriff 1998). The corpus and the senses in its associated dictionary developed

---

[1] http://www.senseval.org

in the HECTOR project were chosen, mainly for its being unseen in the WSD circle, so that no system would have any advantage. For the pilot SENSEVAL in 1998 (or hereafter SENSEVAL-1), 35 English words in various parts-of-speech (POS) were selected, based on word class, frequency, and degree of polysemy.[2] All instances of them (about 200 for each) in the corpus were sampled for the dataset. The corpus was annotated via lexical sense tagging (as opposed to textual tagging), where annotators tagged all instances of the same word in a row, then worked on another word.[3] The validity of the tagging was checked by inter-tagger agreement (ITA). Unlike what would have been expected, ITA turned out to be quite high, reaching 95%. Kilgarriff (1999) believed that the crux was in a clear task definition, qualified annotators, a good sense inventory, and double-tagging with an arbitration phase.

SENSEVAL-1 was run in this order: defining the task; selecting, tagging, and distributing data; participants running their systems; scoring results; and meeting for discussion. Only a lexical sample disambiguation task was piloted. The scoring was done by *precision* and *recall*, defined as follows:

$$Precision = \frac{No.\ of\ correct\ predictions\ made\ by\ system}{No.\ of\ all\ predictions\ made\ by\ system}$$

$$Recall = \frac{No.\ of\ correct\ predictions\ made\ by\ system}{Total\ no.\ of\ items\ in\ evaluation\ data}$$

Participating systems in SENSEVAL-1 varied a lot in different respects. Unsupervised systems were generally outperformed by supervised systems (Kilgarriff and Rosenzweig 1999). The best performing system (JHU) achieved an accuracy of 78.1%. It was a supervised system based on hierarchical decision lists, using a rich set of collocational, morphological and syntactic contextual features with different weightings (Yarowsky 2000). The exercise was agreed to be feasible and worthwhile. More importantly, at least the first step has been taken, with the biggest gap seemingly overcome.

SENSEVAL-2 followed a similar principle and took place in 2001 (Edmonds and Cotton 2001). It has expanded in the task scope and the languages covered. There is the all-words task in which all content words in a sample of texts are to be disambiguated, with data available in English, Czech, Dutch and Estonian. The lexical sample task requires systems to disambiguate all instances of a selection of sample words. Nine languages are covered, including English, Basque, Chinese, Danish, Italian, Japanese, Korean, Spanish and Swedish. The translation task where senses correspond to distinct translations of a word into another language is only held for Japanese. About 35 teams with over 90 systems participated. The sense inventory for the English tasks was based on

---

[2] Evaluation was also done on French and Italian similarly, but in a separate track called ROMANSEVAL.

[3] Such tagging with actual senses is what we call a recognition task, which is presumably more demanding than discrimination but less so than classification.

WordNet 1.7. The hybrid system by Mihalcea (2002), as described in the last chapter, was one of the top performing systems. It achieved 69.8% and 71.2% for coarse-grained scoring in the English all-words task and lexical sample task respectively.

Feeling that simply evaluating WSD by all-words and lexical sample tasks may not further the interests in the field, SENSEVAL-3 in 2004 introduced many new tasks in addition to these traditional disambiguation tasks. In the English lexical sample task, examples were extracted from the British National Corpus. The sense inventory for nouns and adjectives was based on WordNet 1.7.1, while that for verbs was from Wordsmyth. It was already observed that verb senses in WordNet may have been too fine-grained and led to relatively weak system performance in SENSEVAL-2. According to Mihalcea et al. (2004), several of the top performing systems are based on combinations of multiple classifiers, and this shows that voting schemes combining several learning algorithms outperform individual classifiers. The best system achieved 72.9% and 79.3% for fine-grained and coarse-grained scoring respectively. Many of the top systems used kernel methods which are considered state-of-the-art learning algorithms, such as Regularised Least-Squares Classification (Grozea 2004) and Support Vector Machines (SVM) (Strapparava et al. 2004; Lee et al. 2004). The all-words task is apparently more difficult. As reported by Snyder and Palmer (2004), supervised systems scored higher than unsupervised systems in both precision and recall, and the best systems in this track reached the 65–70% range, which was within expectation since the typical inter-annotator agreement was 70–75%.

Palmer et al. (2006) gave a detailed account of various aspects of WSD evaluation up to SENSEVAL-3, and pointed out that there are still issues around the choice of sense inventory being used for evaluation, especially for English. Also, while many researchers still like such in vitro evaluation for its clear task definition, it can hardly lead to new research and it will be as important to demonstrate the positive impact of WSD on NLP applications in the next evaluation exercise.

SENSEVAL-4 was thus renamed as SEMEVAL-2007 to reflect the task variety. Among the 18 tasks, traditional WSD evaluation appeared in many different forms, such as coarse-grained English all-words task, English lexical sample task via English–Chinese parallel text, multilingual Chinese–English lexical sample task, WSD of prepositions, etc. WSD evaluation was also integrated with other semantic analysis tasks or specific applications, such as cross language information retrieval, metonymy resolution, lexical substitution, annotation of affective text, semantic role labelling, etc. SEMEVAL-2010 carried on with this evolution to include many semantic analysis evaluation tasks, including coreference resolution, lexical substitution, keyphrase extraction, argument selection, event detection, amongst others, leaving only a minimal trace of WSD evaluation per se.

3.3 The Old Problems Revisited

Alongside the progress made in the SENSEVAL exercises, a consensus gradually grows in the field regarding the limitation of state-of-the-art WSD systems, and the increasingly pressing need for WSD to go beyond laboratory experiments to clinical applications. Eight years after Ide and Veronis' (1998) recommendations, Agirre and Edmonds (2006) added several more new directions for WSD in their edited volume:

- *Domain- and application-based WSD*—Application-specific research is a major route forward for the field.
- *Unsupervised WSD and cross-lingual approaches*—There is still the need for robust ways to obtain large amount of tagged data with minimum manual effort, to alleviate the knowledge acquisition bottleneck.
- *WSD as an optimization problem*—Instead of handling each word in isolation, the inter-dependencies among senses could be modelled and treated as an optimization problem.
- *Applying deeper linguistic knowledge*—The use of deeper linguistic knowledge rather than better learning algorithms could possibly advance system performance, and this might mean shifting back to knowledge-based methods, but coupled with corpus-based methods.
- Sense discovery—Any static sense inventory will never be able to cover new usages, borrowed words, and the like. Identifying new senses from parallel corpora and the Web would be useful in this regard.

In fact, since WSD has been conceived as a means rather than an end from the very beginning, the value of its research naturally lies on how much a WSD component can improve the overall results of real NLP applications like machine translation and information retrieval. Mihalcea and Pedersen (2005) have also anticipated that in the next five years WSD needs to demonstrate its impact on applications, and it will be deployed in more applications such as web search tools and email organisers, while conventional kind of in vitro evaluation papers (bakeoff evaluation) will meet with more rejection than acceptance.

But what about the old problems mentioned in Ide and Veronis (1998)? Are they really settled and sufficiently understood? In addition to extending our attention to the impact of WSD on other semantic analysis tasks or language applications, are there also other aspects of WSD itself which may have escaped our previous attention but may possibly allow us to surmount the current plateau of system performance? Another look at the lingering issues may bring a new vantage point.

3.3.1 Consensus on Sense Granularity

The fine granularity of the WordNet senses is often blamed as the major obstacle to effective WSD. For instance, it was observed that shifting from the HECTOR to WordNet sense inventory resulted in a substantial drop in performance from

78% to 64% accuracy in the lexical sample task from SENSEVAL-1 to SENS-EVAL-2 (Navigli 2009). This limit posed by the sense inventory on the performance of WSD systems has somehow be held responsible for their uselessness in real NLP applications,[4] and has led to a rethinking of what level of sense granularity is optimal for WSD and more importantly, what is really needed.

The general agreement is toward more coarsely grained senses. On the one hand, WordNet senses are criticised for being unrealistically fine-grained, and sometimes overlapping. On the other hand, even human beings may not easily distinguish such senses. This is where psycholinguistic evidence revitalises in the consideration of the nature of senses in WSD, and the sense distinction issue was thematically discussed in the Workshop on Making Sense of Sense: Bringing Psycholinguistics and Computational Linguistics Together in 2006 (Ide and Fellbaum 2006). Brown (2008) suggested that word senses can be clustered into more general ones as far as WSD is concerned, as there is psycholinguistic evidence indicating that closely related senses may be represented as one sense in the mental lexicon and distantly related senses as discrete entities. Erk et al. (2009) considered the psychology of concepts and suggested to treat sense distinctions in a graded fashion instead of "winner takes all" as traditionally held. We will return with more psycholinguistic evidence for senses in Chap. 4.

Computational linguists have thus reconnected themselves with coarse-grained WSD, clustering fine-grained senses into more general ones (e.g. McCarthy 2006; Navigli 2006b) or using the WordNet lexicographer file labels as supersenses (e.g. Kohomban and Lee 2005). As Ide and Wilks (2006) remarked, NLP applications seem to need homograph level disambiguation. Finer-grained distinctions are rarely needed, though they are likely to remain the concerns for lexicographers. They further recommended that the immediate next steps as far as sense granularity is concerned is to identify a set of inventory-independent sense distinctions at the homograph level which is necessary for WSD, and to enhance stand-alone WSD systems to achieve perfection for homograph disambiguation.

3.3.2 Producing Sense-Tagged Data

Sense-tagged data are essential both for training supervised WSD systems and for evaluating system performance. Large amount of sense-tagged data, with enough examples for individual senses as well as for a wide range of word types, are particularly important for system training. However, manual sense annotation is labour intensive and error-prone. Especially in view of the intractable figure (16 man-years for manually tagging a corpus sufficiently large for broad coverage and

---

[4] With a few exceptions like Chan et al. (2007) and Specia et al. (2006), most have denied the contribution of imperfect WSD components to real NLP applications (e.g. Krovetz and Croft 1992; Sanderson 1994). Plaza et al. (2011), for instance, found a correlation between the overall results of WSD and summarisation of biomedical texts.

high accuracy WSD) projected by Ng (1997), there is obviously a need to either ease the difficulties of manual tagging, or to find cheaper ways to obtain larger amounts of sense-tagged data, perhaps at the expense of quality.

To improve the agreement between annotators in the production of sense-tagged data, Navigli (2006a) introduced a tool called Valido for supporting annotators in assessing the quality of manual and automatic sense assignments. Based on wide-coverage computational lexicons or ontologies like WordNet, the semantic inter-connections among the senses chosen for neighbouring words within the disambiguation context are computed and shown with a semantic graph representation to help annotators visualise and analyse the correctness of the sense tagged for a given word.

McCarthy et al. (2007) mentioned that the first sense or most frequent sense heuristic, apart from being a common back-off strategy in many WSD systems, is itself a powerful heuristic. They pointed out that only 5 of 26 systems in the SENSEVAL-3 English all-words task performed better than the first sense heuristic derived from SEMCOR, and the advantage of such supervised systems, despite their access to hand-tagged training data and use of many contextual features with sophisticated machine learning algorithms, was just marginal. Hence the ability to discover the predominant sense of a word from raw text would help make the first sense heuristic a sound alternative especially in the shortage of manually sense-tagged data for training WSD systems. They therefore collected distributionally similar words (or neighbours) to a target word from parsed data and ranked them according to prevalence to determine the relative predominance of senses. The prevalence scores are based on a combination of distributional similarity score and semantic similarity score of the neighbours to a target word sense. Promising results for all open class POS were reported, especially for nouns with low frequency in SEMCOR.

Supersense tagging, as mentioned above and in the last chapter, is another alternative taken for addressing the data sparseness and sense granularity problem in WSD. For example, Kohomban and Lee (2005) used WordNet lexicographer files as coarse semantic classes and suggested that by learning these generic classes, the limited training data can be reused for the whole class of words without the need for specific training data for individual words.

Chklovski and Mihalcea (2002) took advantage of the ever-growing Web space to collect large amount of training data, sense-tagged by the general public and thus produced at a lower cost than with professional lexicographers. They introduced the Open Mind Word Expert which is a Web-based interface for users to tag words with WordNet senses. It has an active learning component which can help select the most difficult examples to be specifically handled by human taggers. In stage 1, some initial examples are tagged by Web users and these tagged examples are used as training data for identifying the difficult-to-tag cases with active learning. These hard examples are presented to users for tagging in stage 2. This two-stage process enables large volume of tagged data to be collected, especially if the tagging is packaged as a game to the contributors.

Given the call for coarse-grained, realistically distinguishable, and contextually predictable senses, translation differences obtained from parallel corpora naturally provide another means to address the problem, and Diab and Resnik (2002) have shown that unsupervised WSD exploiting translation correspondences in parallel corpora (English-French) as a "radically different source of evidence" gives comparable or even better results when compared to other supervised systems (see description in Chap. 2). This approach has also been subsequently used to generate large amount of noisily annotated training data to train a supervised SVM-based WSD system with features from wide and narrow windows, as well as grammatical relations (Diab 2004). A similar approach was used by Ng et al. (2003) to acquire sense-tagged data from English-Chinese corpora for training a Naïve Bayes classifier with POS, surrounding words and local collocations as features, to disambiguate nouns.

3.3.3 Interaction between Knowledge Sources and Algorithms

The external problems of sense distinctions and availability of tagged data aside, an important question in WSD still remains, and that is the contributions of individual knowledge sources and the effectiveness of various algorithmic approaches. Similar comments have been found in many studies. For example, Dagan and Itai (1994):

> "Further research is therefore needed to compare the relative importance of different information types and to find optimal ways of combining them." (p. 587)

and Ide and Veronis (1998):

> "... to date there has been little systematic study of the contribution of different information types for different types of target words. It is likely that this is a next step in WSD work." (p. 21)

Multiple knowledge sources are used in all state-of-the-art WSD systems without exception. The detailed comparison between five popular machine learning algorithms trained on the same data with a common feature set by Màrquez et al. (2006) suggested that SVM and AdaBoost performed better than kNN, which in turn performed better than Naïve Bayes and Decision Lists. All are superior to the most-frequent-sense baseline though.

However, there is an important unanswered question in this kind of comparison. Although the various popular learning algorithms achieve similar disambiguation results (61–67%), they do not show perfect agreement. It might be unrealistic to expect very high agreement among them given that human annotators also only demonstrate limited agreement most of the time, but this is probably not sufficient to explain the problem away, since all these computational methods did their predictions based on the same set of features. While it is difficult to control how human annotators do the sense tagging (e.g. which knowledge source is held more salient), the fact that methods relying on the same set of disambiguation features only reach limited agreement apparently points to the opacity of the effect of both the knowledge sources and the algorithms.

Similarly, the comprehensive study by Yarowsky and Florian (2002) on relative system performance across different training and data conditions clearly shows the interaction among feature sets, training sizes, and learning algorithms. They concluded that "there is no one-size-fits-all algorithm that excels at each of the diverse challenges in sense disambiguation". For instance, discriminative and aggregative algorithm classes often have complementary regions of effectiveness across numerous parameters, and these results strongly motivate the usage of classifier combination algorithms to incorporate the diverse and unique strengths of these algorithms into a synergistic consensus. This is not only immediately realised in Florian et al. (2002), which investigates and comprehensively evaluates a range of traditional and novel classifier combination algorithms, but also find practical support from the superiority of ensembles of classifiers in SENSEVAL (Mihalcea et al. 2004). The conclusion thus far is that robust combination of diverse classifiers can achieve significant improvement over single classifiers.

Agirre and Stevenson (2006) tabulated from many WSD studies the different knowledge sources available or extracted from various lexical resources and corpora, and their realisation as a variety of features in individual systems. Based on their analysis of the results from these studies, the following conclusions on the effectiveness of different knowledge sources for WSD were drawn (p. 246):

1. All knowledge sources seem to provide useful disambiguation clues.
2. Each POS profits from different knowledge sources, e.g. domain knowledge and topical word association are most useful for disambiguating nouns while local context benefits verbs and adjectives. The combination of all consistently gets the best results across POS categories.
3. Some learning algorithms are better suited to certain knowledge sources; those which use an aggregated approach seem suited to combining the information provided by several knowledge sources. There is also evidence that different grammatical categories may benefit from different learning algorithms.

Although these are concrete and valid observations from empirical testing of various systems using different knowledge sources, the question as to how individual knowledge sources contribute to disambiguation is still left unanswered. Agirre and Stevenson (2006) further suggested that "future work in this area should attempt to identify the best combination of knowledge sources for the WSD problem and methods for combining them effectively", and they tried to attribute the success of the combination of knowledge sources to "that polysemy is a complex phenomenon and that each knowledge source captures some aspect of it but none is enough to describe the semantics of a word in enough detail for disambiguation".

This conclusion and explanation is not really surprising though. McRoy (1992), for instance, already used multiple knowledge sources even in the era of knowledge-based approaches, before large corpora were available for knowledge acquisition. Resnik and Yarowsky (1999) have also pointed out the lexical sensitivity of WSD such that each word requires an individual disambiguator. Leacock et al. (1998) already noted the differential effectiveness of local and topical context for disambiguating nouns and verbs. Specia et al. (2010), using inductive logic programming to learn the rules for disambiguation, found that deep knowledge sources like phrasal verbs managed to contribute to disambiguation even with very few instantiations in the training and test data, whereas selectional restrictions may be more useful for fine-grained than coarse-grained sense distinctions. Also, shallow knowledge sources like bag-of-words, POS tags, bigrams and overlap of definitions are more useful for nouns, while more elaborate knowledge sources like subject-object relations and collocations are more useful for verbs. What remains to trouble us is even though the various learning algorithms have aimed for an optimal combination of different knowledge sources, we nevertheless find limited agreement between systems on their predictions as well as differential performance of the same systems (even the top ones in SENSEVAL or SEMEVAL) on different words. Some attributed this to the differential sense proximity and instance difficulty (e.g. Chugur et al. 2002; Pedersen 2002). Yarowsky (2010) took this to imply that the quality of the feature space can have significantly greater impact on WSD performance than the choice of classification algorithm. The fact in front of us is, unfortunately, that we have not yet been able to sufficiently pinpoint the relation between the contribution of different knowledge sources and the variations of the target words over these years.

3.4 What Have We (Not) Learned So Far?

What are we missing then? Let us consider the positive and negative symptoms of WSD again.

The following are what we can generally and positively say about WSD, based on the observations from many evaluative and comparative studies:

- Popular sense inventories like WordNet are often too fine-grained and WSD systems work better with coarser-grained sense distinctions, which might just be what NLP needs anyway.
- Different kinds of disambiguating information are needed, including syntactic, semantic, and pragmatic knowledge. Systems employing multiple knowledge sources always work better than those relying on a single knowledge source.
- Knowledge-based approaches for WSD often rely on the availability of particular lexical resources containing the necessary disambiguating information, making them more restrictive than machine learning approaches which can scale up simply by generalising patterns from large corpora, whether they are sense-tagged or not.
- Supervised learning systems outperform unsupervised ones in general, assuming that sufficient representative training data is available. Some machine learning

algorithms like kernel-based (e.g. SVM) and example-based methods produce better results than others when trained on the same features.

- There is interaction between machine learning algorithms and knowledge sources. Some algorithms tend to produce better results with topical bag-of-words features, and others tend to perform better when trained with local collocations or syntactic features.
- There is interaction between knowledge sources and the POS of target words. For instance, previous studies have observed that nouns might be more effectively disambiguated with topical cues where verbs and adjectives are more prone to relation-specific information such as selectional preferences.
- Ensembles of classifiers yield better performance than single classifiers across the board. Systems which take into account the predictions given by multiple classifiers based on different learning algorithms or trained on different feature sets to reach a final decision with some voting scheme tend to overcome the weakness of individual classifiers for some reason and produce superior results.

Notwithstanding these positive lessons, current WSD research still may not be able to give an adequate account on the following questions:

- Fine-grained senses might be the artifacts produced by lexicographers, but if their distinctions were somehow based on sufficient contextual and usage differences, why would systems fail to distinguish them? Is this a limitation caused by the lexical knowledge available to a system, the sensitivity of a learning algorithm, or the sense representations within a lexical resource?
- While we find that target words of different POS categories favour different knowledge sources for disambiguation, how can we also account for the intra-POS variation in disambiguation effectiveness with the same and most-favoured knowledge source?
- Is it possible to identify any commonality among the test instances (across different target word types) which can be successfully disambiguated with a certain knowledge source and among those which fail to be disambiguated with the same kind of knowledge?
- What rationales and justifications are there for manipulating the feature sets and combining classifiers in various ways? It has been suggested that different learning algorithms may be biased toward the data in different ways but it has not been transparently explained why they might complement one another.

In particular, if we consider that (1) even the best systems in SENSEVAL do not demonstrate the same level of performance across all test words, and (2) while most learning algorithms aim for an optimal combination of the contextual features, they show limited agreement on the predictions for the same target words even when they use the same feature sets, it follows that there must be something about individual words which is responsible for the apparently different demands on WSD in terms of the most relevant knowledge sources and the best algorithmic approach. Such unknown factors which may have caused the "lexical sensitivity" of WSD must then be beyond external factors like what kinds of knowledge and which learning algorithms are used, and also most likely to be something other than simple classifications like syntactic word classes. They probably have to do with the intrinsic nature of words such that different kinds of words are most effectively disambiguated by some knowledge sources but not others. But then what kind of word classification is relevant to this differential *information susceptibility* is precisely what we need to find out, before further substantial breakthrough is possible for WSD performance. This direction of research, however, cannot be dealt with from the computational perspective alone, and it is therefore time to take a step back to reconsider more evidence from the cognitive aspects of word sense disambiguation by humans, which may provide new insights into the task and open up new ways for it.

3.5 Summary

Despite the many positive lessons learned from WSD evaluation studies in the last decade, there are still unknown factors especially regarding the nature of the target words. Although it is very important to demonstrate the usefulness of WSD in real applications, with the unknowns it seems premature to content ourselves with the plateau seemingly reached by state-of-the-art WSD performance. The missing piece of the puzzle, as we suggested, is possibly not to be found solely from the computational perspective. Thus in the next chapter, let us resort to the cognitive aspects of WSD to look for more clues.

Chapter 4
The Psychology of WSD

Abstract How do humans resolve semantically ambiguous words? It happens that we will not find a direct answer from psycholinguistic studies. Nevertheless, through probing the organisation of words in the mental lexicon and the access of words, particularly those with multiple meanings, in the human mind, useful hints might be found. In this chapter, we focus our attention on the cognitive aspects of word sense disambiguation. We first review the psychological findings on the mental lexicon, including the storage of words, the representation of meanings, and sense distinction. Mechanisms of lexical access will then be discussed, especially with reference to the different factors which might affect semantic activation. Where appropriate, how such psychological models have been realised computationally in automatic word sense disambiguation will be highlighted.

4.1 Words in the Human Mind

Believe it or not, there is literally almost no study in the vast volume of psychology literature which tells us *how* humans disambiguate word senses. Rather, indirect evidence is found within the broader topic of human lexical processing, which mainly deals with the organisation of words and the representation of meanings in the mental lexicon, and the mechanisms of lexical access, especially for words with multiple meanings. Hence in this section, we will expound on these two areas to see what psycholinguistic studies might suggest on the nature of word sense disambiguation (WSD).

O. Y. Kwong, *New Perspectives on Computational and Cognitive Strategies*
for Word Sense Disambiguation, SpringerBriefs in Speech Technology,
DOI: 10.1007/978-1-4614-1320-2_4, © The Author(s) 2013

4.1.1 The Mental Lexicon

Humans know tens of thousands of words, and their language behaviour suggests that the words are systematically organised and efficiently accessed in their internal word repository, often known as the mental lexicon. For a long time, psychologists have hypothesised the mental lexicon as a massive network of inter-connected nodes. As Aitchison (2003) pointed out, the general picture of the mental lexicon so far is one in which there are a variety of links between words, some strong, some weak. Strictly speaking, however, our knowledge of words includes phonological, morphological, syntactic, and even other lower level features like shapes and strokes. Hence network models in different forms and complexity have been proposed under the connectionist roof. For example, McClelland and Rumelhart's (1981) interactive-activation model assumes three levels of processing (feature, letter, and word) which occur simultaneously with excitatory or inhibitory interactions. Others (e.g. Bock and Levelt 1994; Caramazza 1997) suggested that a lexical network should also connect a lemma level and a lexeme level for syntactic and phonological properties respectively, in addition to the conceptual level for semantic relations. Nevertheless, our focus here is entirely on the semantic connections among words, without denying the possible existence of other access levels.

Semantic networks are thus the predominant models for the representation of concepts[1] in the mental lexicon. Collins and Quillian (1969, 1970) proposed a hierarchical network model which is a network of taxonomic and attributive relations. Hence "canary" and "ostrich" are a kind of "bird", which is in turn a kind of "animal". Birds have feathers and can fly, and these attributes are inherited by canaries which have additional characteristics like "can sing", while ostriches are exceptions as they cannot fly. Collins and Quillian used a semantic verification task to test the model, where subjects were presented with sentences in the form of "An A is a B" and were asked to determine whether it is true or false as quickly as possible. The reaction time is then taken to reflect the relative distance between the concepts in the mental lexicon. While such an organisation of concepts is appealing, the major drawback with the model lies in the uniformity of the links assumed between concepts. For instance, the typicality effect was observed, where items which are more typical of a given subordinate take less time to verify. The hierarchical network model thus gave way to the spreading activation model, which retained the network structure, but relieved the strict hierarchical organisation. Hence some nodes are assumed to be more accessible than others and the accessibility is related to factors like typicality and frequency. Collins and Loftus (1975) is the classic study supporting the spreading activation model to human lexical access.

---

[1] The boundary is often fuzzy between concepts and words, and even senses, when they are discussed in the context of semantic memory. For simplicity, we will roughly treat the nodes in a semantic network as corresponding to some lexicalised concepts.

Despite the opacity of the mental lexicon, the psychological reality of the network models receives ample support from experimental psychology. Lexical decision tasks, priming and phoneme monitoring are amongst the most common techniques for studying the mental lexicon. Lexical decision tasks measure a human subject's reaction time in deciding whether a sequence of sound or letters is a word or not, and the results suggest which words are more readily available from the mental lexicon. Semantic priming pre-activates a subject's attention with a prime stimulus, followed by a lexical decision task on the target stimulus. The reaction time is taken to reflect how lexical access might be affected by various factors, tested with different prime-target relations. Such studies support the spreading activation model to human lexical access and offer insights for the organisation and activation of words with different degrees of polysemy and concreteness (e.g. Swinney 1979; Kroll and Merves 1986; Bleasdale 1987; Rodd et al. 2002). They will be further elaborated later in this chapter.

While experiments on lexical access tend to probe the mental lexicon from a speaker's comprehension, the basic and straightforward word association tests can be taken to tap the mental lexicon from one's production. Founded by Galton (1883), this kind of experiment is still in use and it offers first-hand evidence on the nature and strength of associations stored in the mental lexicon. In a word association test, human subjects are usually asked to give the first word which comes to their mind upon hearing or seeing a certain stimulus word. The general patterns of responses from large groups of subjects are often compiled into a set of free association norms. The percentage occupied by a certain response is assumed to indicate the associative strength between the stimulus and that response. Analysis of word association norms reveals that there are several common relations found between the responses and the stimuli, including coordination (e.g. "salt" to "pepper"), superordination (e.g. "colour" to "red"), synonymy (e.g. "hungry" to "starved"), collocation (e.g. "net" to "butterfly"), attributes (e.g. "comfortable" to "sofa") and functions (e.g. "rest" to "chair") (Aitchison 2003; Carroll 2004). Some studies have observed positive correlation between the lengths of the stimuli and the responses, and the association of words of different parts-of-speech. While this approach has been criticised for isolating stimuli from actual context, failing to reflect the variety of links with just one response, and ignoring word frequency, word association tests remain the least intervening method to probe the mental lexicon, and word association norms have valuable applications in many respects (e.g. de Groot 1989). Examples of famous English word association norms include the 1952 Minnesota word association norms (Jenkins 1970) and the Birkbeck word association norms (Moss and Older 1996).[2] Smaller-scale association norms obtained from selected clinical subjects are often used as a psychodiagnostic tool (e.g. Hirsh and Tree 2001).

---

[2] Large-scale word association data are also available for Japanese (Joyce 2005) and German (Schulte im Walde et al. 2008).

4.1.2 Sense Distinction and Meaning Representation

The philosophical issues behind word senses have been raised in Chap. 1. Although senses appearing in dictionaries seem to be the artifacts created by lexicographers, the psychological reality of senses has somehow been observed from humans. How does the mental lexicon compare to dictionaries in terms of sense distinction and granularity? Jorgensen (1990) pointed out that while there is no principled way to distinguish different senses of words, one way to collect data relevant to modelling the lexical memory will be to collect consistent judgements of semantic similarity and difference of use for a large portion of the vocabulary from many speakers of the language. Subjects were asked to sort sentences of some target nouns (chosen from low and high frequency groups, with varying degrees of polysemy) into groups based on their usage or meaning and give a definition to each meaning group. It was found that subjects consistently judged and substantially agreed upon the major senses of most nouns, and more usage samples did not prompt for more number of senses. Hence, to a certain extent, the larger number of senses that lexicographers create may not really be an artifact from larger usage samples. The psychological reality of word senses is also demonstrated in Klein and Murphy (2001, 2002). They pointed out that humans are more aware of the existence of homonyms but are less familiar with the far more prevalent phenomenon of polysemy. Based on the results from sorting and priming tasks, they suggested that related senses of polysemous words like "paper" meaning the newspaper or writing material can simultaneously be dissimilar and are probably represented separately in the mental lexicon, so they are actually treated like homonyms.

We have discussed in Chap. 1 several models for representing senses. How are senses represented in the mental lexicon? We cannot observe how meanings are physically encoded in the neurons, but it is quite unlikely that meanings in the form of definitions are stored as discrete entities in the mental lexicon. For instance, are closely related senses of a word stored under a core sense, or are they stored as separate entities? Pinker (2008) argued that the representation is by *conceptual semantics*, which suggests that word meanings are represented in the mind as assemblies of basic concepts in a language of thought. He compared and contrasted this compositional model with other (less appealing, according to him) alternatives, including:

- *Fodor's Extreme Nativism*, which claims that the meanings of most words exist innately and cannot be decomposed into simpler units, to be triggered by their counterparts in the real world.
- *Radical Pragmatics*, which suggests that the mind does not contain fixed representation of words, as words are fluid and can mean different things in different circumstances. What we draw upon in memory is not a lexicon of definitions but a network of associations among words and the kinds of events and actors they typically convey.

- *Linguistic Determinism*, which holds that language is not a window to human thought but is the language of thought, determining the kinds of thoughts we can think.

To argue for his case, Pinker (2008, p. 150) particularly portrays the incompatibility of the three alternatives as a game of rock-paper-scissors. Differences among language held by Linguistic Determinism contradict the universality of concepts hoisted by Extreme Nativism. The precision of word senses, which Extreme Nativism uses to discredit definitions, casts doubts on Radical Pragmatics, which assumes that one's knowledge of a word is highly malleable. Polysemy which motivates Radical Pragmatics is troublesome for Linguistic Determinism, as it shows that thoughts must be much finer-grained than words.

4.1.3 Influence on Computational Semantic Lexicons

The very notion of association in the network model of the mental lexicon has thus significantly influenced the design of many computational lexical resources. Despite the improved lexical access in modern dictionaries, particularly with their machine readable versions, there is still considerable difference between the mental lexicon and dictionaries. The fuzziness or flexibility tolerated by the mental lexicon can hardly be adequately represented in dictionaries. For instance, humans have the innate ability to handle multiple meanings of words and to deal with the fluidity of these meanings. As Wittgenstein (1958) has argued, two uses of the same word will hardly bear exactly the same meaning because strictly speaking the context in which they are used will never be identical. Humans can nevertheless cope with such vagueness in an apparently effortless way. Moreover, humans seem to be able to organise slightly different meanings on a continuum around a prototype to find the best though possibly imperfect match, conserving the principle of economy. Hence, the treatment of word senses as fixed, discrete and enumerable items in dictionaries soon reveals its inadequacy to account for every single occurrence of a word in text, especially the fluidity resulting from slight variation and innovative usage (e.g. Kilgarriff 1992), which is particularly obvious in the case of metaphorical extensions and systematic polysemy (e.g. Copestake and Briscoe 1995).

The concept of semantic relatedness is much more consequential for lexical resources intended for natural language processing (NLP). Tasks like WSD draw heavily on lexico-semantic information and the relationship among word senses. The influence of the network models proposed by psycholinguists is particularly evident in the semantic processing models designed by computational linguists, such as the semantic network model in Quillian (1968) and Polaroid word model in Hirst (1987) for WSD. It is generally realised that at least some semantic hierarchies or groupings are indispensable (e.g. Chodorow et al. 1985; Calzolari 1988; Vossen et al. 1989; Veronis and Ide 1990; Fontenelle 1997), and there is growing interest in general ontologies as important knowledge sources for NLP, e.g. SUMO (Pease et al. 2002). Investigations on the mental lexicon thus have significantly informed the large-scale lexical resources subsequently produced for NLP use.

WordNet (Miller et al. 1990; Fellbaum 1998) is perhaps the most typical example exemplifying the applicability of psycholinguistic findings in computational linguistics. It is often considered the first large-scale electronic lexical database available, in which word senses are defined by groups of synonyms called synsets and the synsets are linked to each other with respect to various kinds of semantic relations including hypernymy/hyponymy, meronymy/holonymy, antonymy and troponymy. WordNet started out as a psycholinguistic project on network models for the mental lexicon, but turned out to be one of the most popularly used semantic lexicons in computational linguistics, despite its relative weakness in capturing syntagmatic relations. The generative lexicon (Pustejovsky 1991; Pustejovsky and Boguraev 1993), on the other hand, intends to overcome the deficiency of the discrete sense view of conventional dictionaries and represents lexical knowledge based on a set of lexical decomposition principles and a lexical inheritance structure. Word sense ambiguity including the creative use of words is viewed as a dynamic phenomenon, unlike the static view held by conventional sense definitions. This is in concord with the conceptual representation supported by psycholinguists such as Pinker (2008) as discussed above.

One of the first explicit connections between word association norms and computational lexicography is perhaps the work discussed in Church and Hanks (1990). They modelled association strengths with mutual information based on the co-occurrence data in corpora (see also Church et al. 1994). In fact, with the availability of very large corpora, many studies have subsequently tried to simulate the observations from word association norms statistically from large corpora, and such statistical simulation provides concrete and scalable data to enhance computational lexicons. For example, with reference to word association data, Roth and Schulte im Walde (2008) suggested that distributional descriptions based on word co-occurrence obtained from corpora, dictionaries, and encyclopedic resources complement one another, allowing better modelling of word meaning and word similarity.

4.2 Lexical Access and Processing

Assuming that words are stored in a network in the mental lexicon, whether word meanings are directly encoded in the nodes or compositionally represented somewhere, lexical access studies intend to find out how words and meanings are retrieved, especially when multiple meanings might be activated. In particular, psycholinguists are interested in questions like whether sense selection takes place right after the ambiguous word is encountered, or only when sufficient disambiguating context has been presented; and whether all possible senses of an ambiguous word are activated, or only the contextually biased sense is accessed;

and whether the dominant sense will always have an advantage even if the context obviously suggests otherwise, etc. Findings from such research give clues on how humans resolve sense ambiguities.

4.2.1   Lexical Access Models

Different models of lexical access have been proposed to account for how the structural characteristics of words lead to the accessing of the corresponding lexical entries in the human mind. The most influential ones include:

- *The Search Model* (e.g. Taft and Forster 1975), which says that the lexicon is serially scanned until a match is found between the incoming sensory information and a lexical entry.
- *The Logogen Model* (e.g. Morton 1969), which sees each lexical entry as an evidence collecting device or "logogen". A logogen becomes increasingly activated as more features of the incoming stimulus resemble those of the word that it represents. The logogen will "fire" upon a threshold and the word will become available.
- *The Interactive-Activation Model* (e.g. McClelland and Rumelhart 1981), which accesses words like the logogen model but assumes three levels of processing (feature, letter, and word). They occur simultaneously with excitatory or inhibitory interactions.

With specific focus on the semantic aspects, Quillian's (1968) computational model of human semantic memory suggests that our memory stores concepts and meanings of words in the form of a connected network with a mass of nodes interconnected by associative links of different kinds. Such connected information allows us to relate one concept to another, and to draw inference from one event to another, which is the core ability in human reasoning. The association between concepts is found by a method generally known as "spreading activation with marker passing", which also underlies many later WSD programs such as Hirst (1987). The psychological validity of such a network model is evident from subsequent studies on lexical priming and lexical access. Priming studies (e.g. Collins and Loftus 1975) suggest that the processing of a concept (in terms of the reaction time in lexical decision tasks) would be faster if primed by a semantically related concept.

4.2.2   Processing Multiple Meanings

In the event of ambiguity where multiple meanings are possible for a word and disambiguation is needed, the most salient question will be whether some or all of the possible meanings are activated, or just the intended meaning as prescribed by the context is activated.

Lexical access studies have proposed several hypotheses regarding the processing of multiple meanings. Evidence from phoneme monitoring tasks tends to support the *prior choice hypothesis*. For example, Swinney and Hakes (1976) asked subjects to press a button upon hearing a specified phoneme in a sentence, and found no difference in the reaction time whether the target phoneme was immediately preceded by an ambiguous or unambiguous word. However, Foss (1970) seemed to find it the other way. On the other hand, Swinney (1979) experimented with a lexical decision task and found that the reaction time decreased when the string was a word related to some sense of the ambiguous word, even if the context was strongly biased toward another sense. This was true for either the dominant or the secondary sense. Tanenhaus et al. (1979) found similar effect for categorially ambiguous homonyms. These results support the *all-readings hypothesis*, that is, all senses are activated. There are thus no unanimous conclusion (see also e.g. Onifer and Swinney 1981; Seidenberg et al. 1982; Tabossi 1988), but it seems that multiple meanings are activated at least briefly. Hirst (1987) capitalised on such psycholinguistic evidence to develop a lexical disambiguation system having similar properties, that is, activating senses by marker passing with Polaroid words as described in Chap. 2.

More importantly, the influence of context on meaning activation seems to interact with the nature of the individual senses, such as the dominance or familiarity of a sense (that is, frequency), the relatedness among the meanings, as well as concreteness, amongst others.

4.2.3 Frequency

Many studies have found the frequency effect in lexical access. The typicality effect mentioned above, observed from semantic verification experiments with a hierarchical network model of concept organisation, suggests that the familiarity of a concept to a subject tends to affect the reaction time. In other words, more familiar concepts lead to faster lexical access.

The effect of frequency is often investigated by means of a dominant sense and a secondary sense of ambiguous words. Simpson and Burgess (1985) presented homographs out of context and measured facilitation of semantic access at several points after the stimulus from 16 to 750 ms. The dominant sense was found active at all points, whereas the secondary sense became active more slowly. This suggests that the dominant sense may be activated more quickly, but other senses are also activated sooner or later.

Studies on lexical access of ambiguous words have also focused on the effect of different kinds of context on the activation of the dominant and secondary senses of the ambiguous words. Rayner and Frazier (1989) asked subjects to read sentences containing lexically ambiguous words with the disambiguating information appearing at different distances from them, and monitored their eye movements. The ambiguous words were either biased (with a strongly dominant sense) or non-biased. It was found that the integration of a dominant sense with prior context was fast, and that for a subordinate meaning was slow. They thus suggested that a biased word is similar to an unambiguous word as far as the language processing mechanism is concerned.

The frequency effect probably corresponds to the usefulness of the most-frequent-sense heuristic, often employed as back-off strategy and baseline for automatic word sense disambiguation, and is considered a powerful heuristic in itself (e.g. McCarthy et al. 2007). Most statistical classifiers have also taken advantage of the frequency effect, as apparent from the influence of prior probabilities on their predictions.

4.2.4 Polysemy

As far as the nature of the target words is concerned, the relatedness among multiple senses could be another factor. For instance, in the lexical access literature, there is a general finding that visual lexical decisions are faster for words that are semantically ambiguous. Rodd et al. (2002) experimented with several sets of test words more rigorously controlled for sense relatedness and challenged the conventional assumption that the phenomenon of "ambiguity advantage" is a result of ambiguity between multiple, unrelated meanings, instead of multiple related word senses.

As mentioned by Klein and Murphy (2002), the distinction between homonymy and polysemy is not of much concern to most people. They examined the similarity of word senses using categorisation and inference tasks, and found that different polysemous senses of a word were categorised together no more than 20% of the time, which is only slightly more often than different meanings of homonyms. They suggested that polysemy covers a range of related senses, with very close and almost identical meanings on one end and nearly unrelated meanings on the other. The latter are processed like homonyms. This was taken by Ide and Wilks (2006) as support for considering homographs as the realistic sense granularity for WSD.

Klepousniotou (2002) also pointed out "most work in psycholinguistics has concentrated on homonymy, while polysemous words have often been used interchangeably with homographs, homophones, or homonyms to test models and theories of lexical access." She investigated the differential processing of homonymous and polysemous words to see if the distinction between homonymy and polysemy is psychologically real. A cross-modal sentence-priming lexical decision task was used, with variation on four types of ambiguous words (homonymy and three kinds of polysemy). It was reported that the theoretical distinction between homonymy and polysemy was reflected in the results, revealing differential processing depending on the type of ambiguity.

Brown (2008) used priming in a semantic decision task to investigate the effect of different levels of meaning relatedness on language processing. Both reaction time and accuracy varied consistently with four categories of meaning relatedness, suggesting that the distinction between homonymy and polysemy may be a matter of degree and word senses may be represented in the mind as continuous rather than discrete entities.

4.2.5 Concreteness

We have somehow put *concreteness* after *frequency* and *polysemy*, but it is interesting to note that Taft (1991), in the discussion of the *semantic* characteristics of isolated words which may influence lexical processing, listed *concreteness* as the first characteristic, only with *polysemy* and *subjective familiarity* to follow it.

Jorgensen (1990), for the word usage sorting task, suggested that concreteness of a word may increase agreement between judges and concrete words are easier to define. Many psychological studies have also supported that concrete words are easier to learn and understand than abstract words (e.g. Paivio et al. 1968; Kroll and Merves 1986; Schwanenflugel 1991). Studies on lexical access have shown that concrete words and abstract words might be separately stored in the mental lexicon, and the reaction time is often shorter for the former in lexical decision tasks and naming tasks (e.g. Bleasdale 1987; Kroll and Merves 1986). Studies on children's spoken and reading vocabulary have shown that abstract words are acquired later than concrete words during childhood (e.g. Yore and Ollila 1985). Psychologists have also put forth various plausible explanations to account for our differential cognitive behaviour on the two types of words. The dual-coding theory suggests that concrete words are represented simultaneously in a verbal/linguistic system and a non-verbal/imagistic system, while abstract concepts are primarily represented in the verbal system (Paivio 1986). The context availability theory, on the other hand, suggests that the advantage of concrete words comes from their stronger and denser association to contextual knowledge than abstract words (Schwanenflugel 1991). Kousta et al. (2011) pointed out that abstract words are more emotionally valenced than concrete words, and the lack of such experiential information in the two models thus makes them inadequate to account for the representation of abstract words.

While our discussion only focuses on the semantic aspects of the mental lexicon and lexical access, concreteness has been studied in other related areas, demonstrating its importance in human lexical processing. For example, studies have shown that concreteness is a salient lexico-semantic variable which interacts with phonological factors in verbal working memory, where the phonological similarity effect is somehow modulated by whether the word lists to be recalled are abstract or concrete (e.g. Acheson et al. 2010). Neuroimaging evidence from normal and language impaired subjects also supports the differential processing of concrete and abstract concepts in the human mind (e.g. Jefferies et al. 2009).

4.2.6 Implications on Lexical Sensitivity of WSD

The lexical sensitivity of WSD is thus closely related to how meanings are construed and processed by human beings. A variety of factors including imageability, frequency of use, familiarity, etc., are simultaneously shaping the way different concepts are understood, interpreted, and associated. If we categorise words representing different concepts as abstract or concrete, it is apparent that human beings use a range of cognitive strategies to make sense of concepts of different concreteness. For instance, one can understand a "mirror" as the "instrument" which reflects the image of oneself; but the simple hypernymy relation might not be as useful for understanding an "injury". It does not refer to some concrete object, although it could often be visualised (such as a bleeding wound or a broken leg). If one further attempts to describe what "loss" is, the continuum between concrete and abstract concepts is more than obvious.

To most people, concrete concepts are apparently easier to understand and learn than abstract concepts. There is ample evidence from psychology to support this observation as discussed above. The distinction between concrete and abstract concepts is therefore something natural and inherent, which plays an important role in the organisation of our mental lexicon, as well as our lexical processing and understanding mechanisms.

Hence, on the one hand, Quillian's semantic memory model remains influential in our conception of the mental lexicon and has been the foundation for many computational semantic lexicons. On the other hand, from the psychological perspective, we see that the access of multiple meanings and the processing of lexical ambiguity are likely to be influenced by the nature of individual target words. Concreteness would be a typical factor in this regard. It is worthwhile to investigate how the impact of such properties on lexical sensitivity could be modelled in automatic WSD, and whether such modelling could substantially benefit the latter.[3]

Regretfully, however, although psycholinguistic studies test the effect of varied contexts and how they interact with other factors like frequency, sense relatedness, and concreteness, the sentential contexts which researchers manipulate are nevertheless artificial. Very often the disambiguating information is presented with a particular word in the sentence. This word is supposed to strongly suggest the intended meaning of the ambiguous word (or, in the control, the "disambiguating word" may bear different degrees of relatedness to the ambiguous word), and its position in the sentence will be varied to test for the effect of prior and posterior context, as well as the point where sense selection takes place. These studies do not say much, at least not directly, regarding the contributions of different kinds of

---

[3] Yuret and Yatbaz (2010), for instance, have mentioned that the abstract classes were responsible for most of the errors in their supersense tagging with unsupervised method. More in-depth analysis of the concreteness effect will be needed, especially with respect to mainstream supervised WSD.

sentential contexts (and thus knowledge sources) to the disambiguation process. Psycholinguistic studies on lexical access with more careful control on a greater variety of sentential context would be needed to see the interaction between knowledge sources and other properties of the ambiguous words, which could provide more evidence to support the development of automatic WSD systems.

4.3 Summary

As shown in this chapter, the network model plays an important role in human lexical processing, especially the organisation of the mental lexicon and the mechanisms of lexical access. Psycholinguistic evidence has significantly informed the design and construction of computational semantic resources as well as WSD systems, although the influence on the latter has become less obvious, if not minimal, in state-of-the-art supervised systems based on machine learning algorithms. On the other hand, studies on lexical access with control on the various characteristics of words provide inspiring information for us to reconsider the lexical sensitivity issue in WSD. To this end, we will focus on concreteness for further investigation.

Chapter 5
Sense Concreteness and Lexical Activation

Abstract Psycholinguistic evidence has thus suggested the differential processing of concrete and abstract concepts by the human mind. This chapter further explores the mental lexicon with respect to the concreteness and abstractness of concepts based on word association data. Since lexical resources including computational semantic lexicons play a critical role in automatic word sense disambiguation, we aim at investigating to what extent such concreteness distinction is modelled in existing lexical resources. It was observed that concrete and abstract noun senses tend to exhibit consistently different lexical activation patterns, and the results suggest that sense concreteness may serve as a possible alternative classification of word senses relevant to the lexical sensitivity of word sense disambiguation, as well as to the contributions of different knowledge sources in the task.

5.1 Computational and Mental Lexicons: Mutual Interaction

In the previous chapters, we have established several key points regarding the nature of word sense disambiguation (WSD) from the computational and cognitive perspectives:

- Multiple knowledge sources for characterising word senses and subsequently disambiguating new occurrences play a critical role in automatic WSD.
- The mental lexicon has long been hypothesised as a massive network connecting concepts with different kinds of relational links.
- Such network models have inspired the design of computational semantic lexicons and their deployment in automatic WSD systems.
- Concrete and abstract concepts are apparently stored and processed by different mechanisms in the human mind.

O. Y. Kwong, *New Perspectives on Computational and Cognitive Strategies for Word Sense Disambiguation*, SpringerBriefs in Speech Technology, DOI: 10.1007/978-1-4614-1320-2_5, © The Author(s) 2013

Psychologists often use word association tests to probe the organisation of our mental lexicon. The lexical association data, in the form of word association norms, serve as a bridge between the internal evidence and the external modelling, and often inform the construction of semantic lexicons for computational purposes. WSD systems will in turn make use of such computational lexicons or acquire similar information from large corpora to do the task. For instance, psychologists may tell us that high-imageability words might be more information-loaded with strong links departing from them in our mental lexicon (e.g. de Groot 1989); while computational linguists might exploit the class information and node density in semantic lexicons for WSD (e.g. Resnik 1995b; Agirre and Rigau 1996). The interaction and mutual advancement between internal cognitive evidence and external computational modelling thus form an important feedback loop in the understanding of WSD mechanisms.

In this chapter, we take a closer look at this feedback loop by summarising our earlier studies on the analysis of lexical association patterns (Kwong 2007, 2009). In particular, the lexical associations triggered in word association tests were analysed against existing computational lexical resources to probe the impact of sense concreteness on semantic activation. Concreteness, however, is difficult to define.[1] Psychologists (e.g. Kroll and Merves 1986) defined it in terms of im- ageability, but imageability is also a highly subjective measure, and the con- creteness of a concept also comprises many other dimensions. For an operational definition, we make use of the classifications in existing knowledge sources commonly used in natural language processing (NLP) studies, including WordNet (Fellbaum 1998) and SUMO (Pease et al. 2002).

As reviewed in Chaps. 2 and 3, knowledge-based WSD methods address the need for multiple knowledge sources by using semantic networks containing different kinds of semantic relations (e.g. IS-A, PART-OF, thematic relatedness, etc.) and/or extracting such information from large corpora, whereas machine learning methods address the issue by learning an optimal combination of the various knowledge sources for disambiguating individual target words. Although many comparative studies have revealed the complex interaction between knowledge sources and learning algorithms, a comprehensive qualitative and objective account for the lexical sensitivity of the task is still lacking.

To say that different information types contribute variably to different target words is essentially presupposing that different types of lexical information vary in their effectiveness to characterise a sense and distinguish it from other senses of the same word. Thus it is not enough to just conceptualise senses by a certain dimension (e.g. a certain semantic relation) across the board. The very intrinsic nature of a given word or sense with respect to different semantic dimensions must also be thoroughly examined.

---

[1] We use *concreteness* as a generic term to refer to a continuum between highly concrete and highly abstract concepts.

Leacock et al. (1998), for example, observed that "the benefits of adding topical to local context alone depend on syntactic category as well as on the characteristics of the individual word". Such "characteristics" are equivalent to what we call the *intrinsic properties* or the *nature* of the target words (as mentioned in Chap. 1), which might include concreteness, frequency, sense relatedness, part-of-speech, amongst others, all of which are critical for understanding the lexical sensitivity of WSD.

Here we focus on one aspect of the intrinsic nature of words, namely *sense concreteness*, and explore how it relates to lexical association in our mental lexicon and how it might affect the effectiveness of various kinds of semantic knowledge in disambiguation. Hence, on the one hand, we investigate the effect of sense concreteness as one aspect of the intrinsic nature of words on semantic activation and association responses; on the other hand, we study the implications of such differential semantic association patterns, if any, on the organisation and modelling of our mental lexicon, as well as the lexical sensitivity of WSD.

5.2 Analysing Word Association Responses

The word association responses in Hirsh and Tree (2001) were compared against WordNet, a widely used semantic lexicon, and SUMO, a popular ontology. Since word association is commonly used to probe the organisation of the mental lexicon, and computational lexicons and ontologies are assumed to model human conceptual structure, the comparison is expected to allow us to better understand the human semantic repertoire and thus the information demand for individual words in the lexically sensitive disambiguation process.

5.2.1 Research Questions

The following research questions are particularly salient to the organisation of human semantic memory and their proper computational modelling:

- If a word has multiple meanings, which meaning or meanings will most likely be activated?
- Does sense concreteness matter? For example, are concrete senses more readily activated than abstract senses?
- Do frequencies of the individual senses matter?
- Do the type and strength of association depend on sense concreteness? Do abstract/concrete senses tend to be more closely and strongly associated with other abstract senses or concrete senses?

Table 5.1 Examples of word association responses from Hirsh and Tree (2001)

Stimulus word	Response words (Associative frequency)
Beer	Drink (0.24), Lager (0.11), Pint (0.11), Wine (0.11), Glass (0.07)
Cheese	Wine (0.13), Biscuit (0.07), Board (0.07), Bread (0.07), Cheddar (0.07)
Circus	Clown (0.42), Acrobat (0.13), Act (0.09), Tent (0.07), Performance (0.04)

Table 5.2 WordNet lexicographer files for noun senses

No.	Name	Contents
03	noun.Tops	Unique beginner for nouns
04	noun.act	Nouns denoting acts or actions
05	noun.animal	Nouns denoting animals
06	noun.artifact	Nouns denoting man-made objects
07	noun.attribute	Nouns denoting attributes of people and objects
08	noun.body	Nouns denoting body parts
09	noun.cognition	Nouns denoting cognitive processes and contents
10	noun.communication	Nouns denoting communicative processes and contents
11	noun.event	Nouns denoting natural events
12	noun.feeling	Nouns denoting feelings and emotions
13	noun.food	Nouns denoting foods and drinks
14	noun.group	Nouns denoting groupings of people or objects
15	noun.location	Nouns denoting spatial position
16	noun.motive	Nouns denoting goals
17	noun.object	Nouns denoting natural objects (not man-made)
18	noun.person	Nouns denoting people
19	noun.phenomenon	Nouns denoting natural phenomena
20	noun.plant	Nouns denoting plants
21	noun.possession	Nouns denoting possession and transfer of possession
22	noun.process	Nouns denoting natural processes
23	noun.quantity	Nouns denoting quantities and units of measure
24	noun.relation	Nouns denoting relations between people/things/ideas
25	noun.shape	Nouns denoting two and three dimensional shapes
26	noun.state	Nouns denoting stable states of affairs
27	noun.substance	Nouns denoting substances
28	noun.time	Nouns denoting time and temporal relations

5.2.2 Data and Resources

5.2.2.1 Word Association Norms

The 90 stimulus words in Hirsh and Tree's (2001) word association test were used as target words. Hirsh and Tree analysed the difference between the responses from young adults and those from older adults, and we only considered the top five responses elicited from the young cohort. Table 5.1 gives some examples of the association responses.

One shortcoming of word association norms is its lack of sense distinction. It is not obvious on the surface which sense of a polysemous stimulus (or response) word is actually activated. Given the relatively small amount of data, we analyse the association responses with manual distinction at the sense level with respect to some existing sense inventory. For analysis at a larger scale, automatic sense clustering like the soft cluster analysis in Melinger et al. (2006) could be used.

5.2.2.2 WordNet

WordNet (Fellbaum 1998) organises word senses in the form of synsets with relational pointers linking among different synsets to form some sort of a semantic hierarchy. Senses are also organised under 45 lexicographer files based on syntactic category and logical groupings, 26 of which are relevant to noun senses, as shown in Table 5.2.[2] WordNet 3.0 is used here as:

- a word sense inventory,
- a computational model of the mental lexicon in the form of a semantic network, despite its known bias toward paradigmatic relationship in general, and
- a means to distinguish between concrete and abstract concepts.

5.2.2.3 SUMO: Suggested Upper Merged Ontology

SUMO is a large, open source, formal ontology stated in first-order logic (Pease et al. 2002). Logic and inference mechanisms are applied to the conceptual level to model human reasoning mechanisms. According to Niles and Pease (2001), "SUMO was created by merging publicly available ontological content into a single, comprehensive and cohesive structure". Synsets in WordNet 3.0 have been mapped to SUMO concept nodes, hence making the correspondence between concepts and lexical items available for a wide range of language processing applications. The root node for SUMO is *Entity*, which subsumes *Physical* and *Abstract*. For the present purpose, we make use of this dichotomous categorisation with a simplified third level of concept nodes, as shown in Fig. 5.1.

5.2.3 Operational Definition of Sense Concreteness

Concept concreteness is somehow not easy to define and may depend on multi-dimensional factors. We tried different operational definitions based on WordNet and SUMO, to reduce bias and evaluate the usefulness of both resources.

---

[2] Excerpt from http://wordnet.princeton.edu/wordnet/man/lexnames.5WN.html

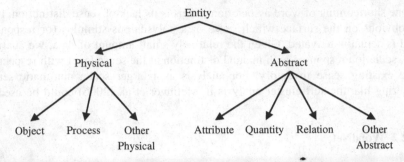

Fig. 5.1 SUMO top level concepts with modification

Among the 26 noun-related lexicographer files in WordNet, we identified 7 concrete classes and 19 abstract classes. The concrete classes include *noun.animal*, *noun.artifact*, *noun.body*, *noun.food*, *noun.object*, *noun.person*, and *noun.plant*. The rest are considered abstract classes, including *noun.attribute*, *noun.cognition*, *noun.feeling*, *noun.motive*, etc.

With SUMO, concreteness is defined according to the *Physical* and *Abstract* nodes in the ontology. Concepts under the *Physical* branch are considered concrete, and those under the *Abstract* branch are classified as abstract. WordNet synsets have been mapped to the concept nodes in SUMO. Hence, for example, a synset mapped to the *FruitOrVegetable* node subsumed by *Physical* is considered concrete. Similarly, a synset mapped to the *SubjectiveAssessmentAttribute* node subsumed by *Abstract* is considered abstract.[3]

5.2.4 Association Patterns

5.2.4.1 With Concreteness Defined by WordNet

Among the 90 target words, 17 are monosemous and 72 are polysemous according to the WordNet noun database (excluding proper noun usages). One target word ("sly") was found only in the adjective database in WordNet, and was ignored in all subsequent analysis. The polysemous words have 2 to 15 senses, with an average of 4.6 senses. Only 33 words have solely noun senses, and the others are categorially ambiguous too.

---

[3] The hierarchical chain between *FruitOrVegetable* and *Physical* in SUMO is: *FruitOrVegetable* < *ReproductiveBody* < *BodyPart* < *OrganicObject* < *CorpuscularObject* < *SelfConnectedObject* < *Object* < *Physical*, where *A* < *B* means *A* is subsumed under a more general concept *B*. The chain linking *SubjectiveAssessmentAttribute* and *Abstract* is: *SubjectiveAssessmentAttribute* < *NormativeAttribute* < *RelationalAttribute* < *Attribute* < *Abstract*.

Table 5.3 Association patterns w.r.t. WordNet relations

Word type	Sense concreteness	Overlapping			
		WNAsso1 Only	WNAsso2 Only	Both	None
Monosemous	Concrete (16)	1	6	8	1
	Abstract (1)	0	0	0	1
Polysemous	All Concrete (20)	2	5	11	2
	All Abstract (3)	0	1	1	1
	Both C & A (49)	0	22	27	0

Only 1 (out of 17) monosemous word is abstract, and the other 16 are concrete. For the polysemous words, 20 (out of 72) only have concrete senses, 3 only have abstract senses, and 49 have both concrete and abstract senses. Altogether there are 216 concrete senses and 136 abstract senses. As Hirsh and Tree (2001) had indicated that their stimuli were "mostly names of concrete or picturable objects or likely to elicit the name of a concrete object", it is not surprising that concrete senses are more abundant. However, they did not mention with respect to which sense(s) the "picturability" was determined in the event of polysemy, and this is what we are going to further explore.

The stimulus–response pairs from the word association data were analysed with respect to two groups of word associations obtained from WordNet. The first (*WNAsso1*) consists of all *words* in the synsets (words composing the synsets only, excluding glosses and examples) directly related to the synset(s) to which the target word (that is, stimulus) belongs. These directly related synsets include antonyms, hypernyms, hyponyms, holonyms, meronyms, and coordinate terms. The second (*WNAsso2*) consists of the words in the glosses and examples in these related synsets. Thus *WNAsso1* could be taken as the cluster of words corresponding to mostly narrow paradigmatic relations with a target word, and *WNAsso2* more likely to be broader associations, including syntagmatic relations.

Table 5.3 shows the number of target words found for the various overlapping scenarios, where the overlapping could correspond to one or more senses of a given target word. For all sense types, the "WNAsso2 Only" and "Both" columns make up the majority, and only three cases overlap with purely paradigmatic responses.[4] This is in consensus with Hirsh and Tree's analysis. There are, however, a few exceptional cases which have none of their responses overlapping with any of our WordNet relations.

As mentioned in Chap. 4, the lexical access literature suggests that multiple senses might be at least briefly activated in the case of polysemy. However, the concreteness factor has not been systematically explored in this regard. For 21 out of the 49 polysemous words with both concrete and abstract senses, the stimulus–response relations falling under *WNAsso1* or *WNAsso2* correspond to one or more

---

[4] But this does not preclude any broader associations or syntagmatic responses which might not be detected with the limited associations obtained from WordNet.

of their *concrete senses only*. This is in sharp contrast to 6 (out of 49) polysemous words with stimulus–response relations corresponding to their *abstract senses only*. This observation thus suggests that in the case of polysemy with both concrete and abstract senses, the concrete concepts are apparently more accessible from the mental lexicon, assuming word association responses reflect the closest and strongest associations in it.

It is also observed that association responses are seldom entirely paradigmatic. This is not surprising as broader associations including syntagmatic associations provide many more possibilities. Nevertheless, about 38% of all target words have their stimulus–response pairs falling under *WNAsso2 only*. Such absolute dominance of broader associations may account for the usefulness of topical knowledge sources for automatic WSD in general. However, the apparent inferiority of paradigmatic responses might also be an artifact of the WordNet classification itself. For instance, the hypernym of "ankle" is "gliding joint", and that for "kennel" is "outbuilding", which are obviously too specialised for daily usages and conception. Hence, even though in the word associations the top response for "ankle" is "foot", they are not straightforwardly related in the WordNet noun hierarchy.

5.2.4.2 With Concreteness Defined by SUMO

Hirsh and Tree's human subjects might not have always responded to a stimulus word as a noun, as they were not specifically instructed to do so. Thus we narrowed our focus to the 33 target words which *only* have noun senses in WordNet, resulting in 13 monosemous and 20 polysemous nouns. The number of senses for the polysemous words ranges from 2 to 8, averaging at 4.25. According to our SUMO definition of concreteness, 11 of the monosemous words are concrete and 2 are abstract. Among the polysemous words, 10 only have concrete senses, and the other 10 have both concrete and abstract senses. This distribution is somehow slightly different from that with concreteness defined by WordNet.

The concreteness definitions by WordNet and SUMO may not always be in concord. For example, it is observed that senses denoting processes often belong to *noun.act* or *noun.process* in WordNet, which we classified as abstract; but *Process* is under *Physical* in SUMO, which becomes concrete according to our definition. Notwithstanding such difference, SUMO offers us another dimension to define sense concreteness and analyse the association responses.

Among the 10 polysemous words which have both abstract and concrete senses, 1 elicited association responses from its abstract senses only, 6 from their concrete senses only, and 3 from both kinds. This reinforces our earlier observation from WordNet that concrete senses are more readily activated than their abstract counterparts.

As for the analysis of the stimulus–response relations, the results are shown in Table 5.4. Three types of association response were classified:

Table 5.4 Association patterns w.r.t. SUMO relations

Stim\Res	Obj	Proc	OP	Attr	Rel	Qty	OA	UC	Total
Obj	98	5	0	10	2	6	0	13	134
Proc	2	1	0	2	0	0	0	1	6
OP	0	0	0	0	1	0	0	1	2
Attr	8	1	0	1	0	0	0	2	12
Rel	5	1	0	0	2	0	0	1	9
Qty	2	0	0	0	0	0	0	0	2
OA	0	0	0	0	0	0	0	0	0
Total	115	8	0	13	5	6	0	18	165

Obj = Object, Proc = Process, OP = Other Physical, Attr = Attribute, Rel = Relation,
Qty = Quantity, OA = Other abstract, UC = Unclassified

- both the stimulus sense and the response sense are from the same SUMO high level concept,
- the stimulus sense and the response sense are from different SUMO high level concepts, and
- response in another part-of-speech or unclassified.

We mentioned in the previous analysis with WordNet that pure narrow paradigmatic responses are rare. In the current analysis, however, the results suggest that the closeness and strength of association between paradigmatically related senses might somehow depend on the concepts concerned. On the one hand, the concrete senses in this analysis are mostly under *Object*. Responses to an *Object* are also predominantly under *Object*. On the other hand, it is generally realised that the links in computational semantic networks, like those of WordNet and SUMO, are by no means uniform links. The depth of a node within a hierarchy and the density within a certain branch in the hierarchy may reflect the variation in association strength and distance. Hence concepts might be more closely and strongly associated if they are deeper down in a hierarchy or more densely located. Apart from this principle, however, some concepts just appear to be more closely and strongly associated, where narrow paradigmatic responses are more likely to be seen. For instance, the monosemous stimulus word "puma" is mapped to the concept *Feline*, and many of its responses such as "cat", "leopard", "lion" and "tiger" are also sense-mapped to *Feline*. It is a similar case for stimulus senses mapped to concepts like *FruitOrVegetable*. Other concrete concepts, like *Device*, may not elicit as many responses under *Device*, but might elicit responses corresponding to more various concepts like *Substance*. Abstract senses, unlike concrete senses, tend to elicit responses under a completely different branch. For example, senses under *Quantity* seldom lead to responses under *Quantity*, but often elicit responses under *Object* instead.

5.2.5 Implications on WSD

We can therefore summarise from the above observation and analysis the following points:

- Where a polysemous word has both concrete and abstract senses, concrete senses are more readily activated.
- Concrete senses tend to elicit concrete association responses, whereas abstract senses are more likely to lead to more various responses.
- Broad associations are more common than pure paradigmatic associations, which are nevertheless primarily modelled in existing computational lexicons and ontologies.
- The closeness and strength of the links between concepts do not only depend on the depth and density of a hierarchical network, but probably also on the particular concept in question.
- Human association responses comprise interacting multi-dimensional relationship, which is hardly adequately modelled at present.

These results bear important implications for NLP and WSD, especially regarding the lexical sensitivity of WSD and the classification of senses in computational lexicons for WSD. First, although frequency and familiarity might be a co-factor, concrete concepts seem to be more easily activated in the mental lexicon, and even in the case of polysemy, concrete senses appear to be more accessible than abstract senses. Abstract senses, in practice, might have lower frequency and at the same time lack distinguishing features. Statistical and supervised methods in general rely on prior sense distributions and probabilities of contextual features, often leaving abstract senses harder to disambiguate. More work on the association of abstract words and senses are needed to enhance our understanding of the semantic memory and to improve the automatic disambiguation of words used in their abstract senses. Second, while narrow paradigmatic associations form an important part of our semantic knowledge, the observed dominance of broad associations including syntagmatic associations might inform the computational modelling of the mental lexicon, such that different weights might be attached to different kinds of associations for words with different nature. Moreover, with respect to the organisation of computational semantic networks, our analysis further suggests that it might not be sufficient to render lexical semantic relations in any single dimension. There are various dimensions interacting with one another, and automatic WSD is a lexically sensitive task, requiring a variety of knowledge sources combined in different ways. Computational semantic lexicons should be enriched by taking various dimensions into account for organising the lexicon and depicting their lexico-semantic relationship. Hence it is worth to investigate the feasibility of enriching existing lexical resources like WordNet with possibly an alternative classification of word senses based on the intrinsic nature of words, such as concreteness, in addition to conventional conceptual classifications.

5.3 Concreteness as a Knowledge Source

Hence the distinction between concrete and abstract concepts is something natural and inherent, which plays an important role in the organisation of our mental lexicon, as well as our lexical processing and understanding mechanisms. By analogy, if this common-sense phenomenon can be properly modelled in computational semantic lexicons, it should benefit NLP tasks like WSD which typically draw heavily on lexico-semantic information. However, few studies have addressed how this intrinsic property of word senses might imply on the construction of computational semantic lexicons and its potential function as a knowledge source for WSD. For example, how can concreteness be included in common ontological organisation of semantic lexicons? If concrete concepts are easier than abstract ones for most people, and polysemy has been shown to affect lexical access, should we expect concrete senses to be more easily disambiguated than abstract senses? If this is the case, should we take it into account when evaluating disambiguation performance, perhaps in addition to the similarity among senses (e.g. Resnik and Yarowsky 1999; Palmer et al. 2006)? Should different information be employed for disambiguating concrete and abstract senses?

One major obstacle to further studying the impact of concreteness on WSD is, ironically, concreteness is not concrete itself. In reality, concreteness is a matter of degree and largely subjective judgement, often affected by many factors like physical existence, familiarity, imageability, frequency of occurrence, etc. In psychology studies, concreteness has often been measured by averaging human ratings on a sample of words. Human raters give a score for each word on a scale (e.g. from 1 for highly abstract to 7 for highly concrete), often depending on the familiarity and imageability of the concepts, and frequency of occurrence, but more heavily on their own experience and probably private cognition. This way of measuring concreteness is limited in scalability. In addition, ratings on word samples do not say anything on the effect of polysemy. A concrete word may also have some abstract senses, and vice versa, and the raters may not be thinking of the same sense when rating a word. It is not always clear which sense or senses make up one's overall impression of a word.

Only if we can have a way to objectively measure concreteness will it be possible for us to pursue further studies on the role of such intrinsic nature of word senses in lexical resources and natural language processing. We therefore investigate the feasibility of obtaining such a measure systematically and objectively. To this end, we explore the potential use of dictionary definitions and summarise the method proposed in Kwong (2011).

5.3.1 Definition Styles and Surface Patterns

The way we propose to measure concreteness is based on a simple intuition. Dictionaries constitute an important source of our lexical knowledge. Language learners often form their perception and understanding of words from dictionary definitions. Professional lexicographers are trained to write definitions informatively and consistently, and it is generally assumed in lexicography that concepts corresponding to tangible objects or intangible things are more appropriately defined by different styles. Dictionary definitions thus reflect how lexicographers (who are also human beings) perceive the concepts being defined.

The way in which definitions are structured and phrased has undergone considerable evolution in modern lexicography, from "lexicographese" to more natural prose, and even full-sentence definitions (Atkins and Rundell 2008). A common way to define a concept is by means of genus (superordinate concept) and differentiae (distinctive features), such as "bag" is defined as a kind of "container" distinguished from other containers by being flexible and having a single opening. For words which are not easy to be defined by a genus term, the definition is often composed with a synonym, a collection of synonyms, or a synonymous phrase. Sometimes it may not be easy to isolate the sufficient and necessary conditions for a sense, and lexicographers will capture the essential constituents in the form of prototype. This is usually combined with genus and differentiae but additionally specifies what is typical of a referent with words like "typically" or "usually". For the rest, where a referent is unlikely to be available, meanings will be explained through their usage in real text. While tangible objects and physical actions are more easily defined with genus, differentiae and prototype, abstract concepts as well as other aspects of meanings like connotation and collocations often need to be defined by other means. We thus focus on the surface structure of the various traditional defining styles, exploiting the regularity exhibited therein, for some objective measures of concreteness.

A scoring system was implemented by mapping the structure of a definition given by a dependency parser to a series of defining patterns roughly corresponding to various degrees of concreteness (Kwong 2008a). The basic assumption is that the more concrete a concept, the more conveniently and convincingly it can be explained with reference to its superordinate concept and distinguishing features. The dependency parser from Lund University (Johansson and Nugues 2008) was used, and the presence or absence of certain dependency relations were detected from the parse results. Only those parses consisting of a root (ROOT) and a predicative complement (PRD) in the form of a noun were processed. Any of the following dependency relations on the PRD would be considered differentiae: Apposition (APPO), Location (LOC), Modifier of nominal (NMOD), Object complement (OPRD), and Temporal (TMP). If any of such relations has a dependency from words like "usually", "often", "typically", etc., it would be considered a prototype.

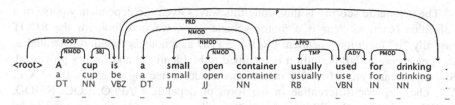

Fig. 5.2 Example of dependency parse by the Lund University Parser

Let Genus X = noun at PRD	
Let Score = 4	
if (X...of...Y) is found, **then**	
if X ∈ {kind, type, etc.} and Y is mass	// e.g. a kind of commercial enterprise ...
then Score = Score - 1	
else if X ∈ {group, part, etc.} and Y is mass	// e.g. a series of related events ...
then Score = Score - 1	
else if Y is –ing verb	// e.g. the manner of speaking to ...
then Score = Score - 2	
else	// e.g. the content of cognition ...
if X is mass **then** Score = Score - 1	
if Y is mass **then** Score = Score - 1	
stop	
if X ∈ {person, someone, anyone, etc.}	// e.g. someone who controls resources ...
then Score = Score + 1, **stop**	
if X ∈ {something, somewhere, etc.}	// e.g. something intended as a guide ...
then stop	
if ∃ Dependency D ∈ {NMOD, APPO, LOC, TMP,	// prototype
OPRD} ← X at any word after X **and** D is preceded	// e.g. ... usually used for drinking
by {usually, typically, often, etc.}	
then Score = Score + 1	
if ∃ Dependency D ∈ {NMOD, APPO, LOC, TMP,	// differentiae
OPRD} ← X at any word after X	// e.g. [a motor vehicle] with four wheels
then Score = Score + 1	
if no prototype **and** no differentiae **and** X is mass	// synonymous phrase with mass noun,
then Score = Score - 1	// e.g. brilliant radiant beauty
else if ∃ Dependency D ∈ {NMOD} ← X at any word	// pre-modifiers,
before X	// e.g. a very young mammal
then Score = Score + 1	
if no prototype **and** no differentiae **and** no pre-modifier	// minimal NP treated as synonym
then Score = Score - 1	// e.g. an idea

Fig. 5.3 Flow for concreteness scoring from dependency parses

The automatic scoring is thus done this way: Given the dependency parse of a definition (such as "car is a motor vehicle ..."), we first look for the ROOT (usually "is") and its dependent PRD. If a PRD as a noun is found, it is treated as the genus and an NP definition is assumed to be identified and further processed. Subsequent analysis of an NP definition includes detecting genitives like "X of Y", checking for differentiae in the form of dependent APPO, LOC, NMOD, OPRD, or TMP to the PRD, and prototypes if such dependencies are marked by words like "usually" or "typically". If no differentiae, prototype, or pre-modifier is present, the definition will be considered a synonymous phrase. The genus is also checked if it is possibly an abstract or mass noun, which is currently approximated by the absence of any indefinite articles and plural markers. An example of parse results is shown in Fig. 5.2 and the scoring flow is summarised in Fig. 5.3 in the form of pseudocodes with examples.

5.3.2 Data Sources and Procedures

Kroll and Merves (1986) used 100 abstract and 100 concrete nouns in their lexical study. The words were rated by human subjects for concreteness on a 7-point scale. The abstract and concrete words were matched on the basis of word frequency and word length. The word frequency data were taken from Kucera and Francis (1967). A total of 100 word samples (50 concrete and 50 abstract) with frequency greater than 20 were selected from Kroll and Merves' list. Sense definitions were collected from WordNet 3.0 glosses. For our word samples, the concrete words have 1 to 17 senses and the abstract words have 1 to 9 senses, with 4.36 and 3.44 senses on average respectively.

Since the average human ratings in Kroll and Merves (1986) are only for the word level concreteness, four human judges were asked to rate the selected words and their individual senses on a 7-point scale (with 7 for highly concrete and 1 for highly abstract).

The WordNet definitions were first made complete sentences by preceding them with the headword as subject such as "car is a motor vehicle...", and then parsed by the Lund University Parser. The parsing results were analysed by the scoring program. The scores for individual senses were compared to the average ratings by our four human judges. For the overall concreteness of a word, we tried two ways for estimating it from the individual senses. One is to take the average of the scores from all senses (AvgDef), assuming all component senses of a word contribute equally to its overall concreteness; and the other is to take the score for the first sense (FirstDef), assuming the most frequent or familiar sense dominates one's perception of the word in general. These two measures were compared with the ratings in Kroll and Merves' study and those given by our human judges.

Table 5.5 Summary of concreteness ratings at the word level

	K&M	AvgR	AvgDef	FirstDef
All word samples (N = 100)				
Mean	4.27	4.58	4.34	4.36
SD	1.76	1.79	0.92	1.37
Abstract samples (N = 50)				
Mean	2.63	2.96	4.11	3.88
SD	0.58	0.73	1.03	1.47
Concrete samples (N = 50)				
Mean	5.92	6.19	4.58	4.84
SD	0.63	0.77	0.74	1.08

5.3.3 Concreteness Ratings from Dictionary Definitions

Given the psycholinguistic evidence on the validity of the concrete/abstract distinction, we first see if the two groups of words result in significantly different scores as estimated by different ways (e.g. AvgDef and FirstDef) for the word-level concreteness. Table 5.5 shows the mean and standard deviation of the concreteness ratings for the 100 word samples (and for the 50 abstract and 50 concrete samples separately) obtained from human raters and WordNet definitions. The column K&M refers to the original rating from Kroll and Merves (1986). AvgR refers to the average of our four human judges. AvgDef and FirstDef refer to the scores thus obtained from WordNet definitions.

The distinction between the two groups of words is particularly clear from the human ratings. With K&M, which is our data source, the abstract samples have a mean rating of 2.63 with standard deviation of 0.58, and the concrete samples have a mean rating of 5.92 with standard deviation of 0.63. A similarly apparent distinction is also seen in AvgR, with mean at 2.96 and standard deviation at 0.73 for the abstract words, and mean at 6.19 and standard deviation at 0.77 for concrete words. Interestingly, our human judges seem to be more lenient in their scores than those in K&M, as shown by the higher mean ratings for AvgR in Table 5.5. The distinction between concrete and abstract words is nevertheless apparent among human judges.

The difference between the mean scores for abstract words and concrete words is much smaller with the estimation from definition scores. The mean is above 3 for abstract words, and below 5 for concrete words, with both AvgDef and FirstDef. In fact, there is a mode at 5 for concrete words, and the majority did fall on the high side above 4. The scores estimated from WordNet definitions by FirstDef are roughly bi-modal for the abstract words, while those by AvgDef are somehow averaged out giving a mode at 4. This suggests that words deemed abstract often possess considerable concrete senses, and the first senses of about half of the abstract words are not necessarily abstract or are not defined as if they are abstract in WordNet.

It turns out that the first sense of a word in WordNet seems to be a relatively better indicator of the overall word concreteness than the average of all senses. It

Table 5.6 Summary of
concreteness ratings at the
sense level

	AvgR	WN
All sense samples (N = 390)		
Mean	4.81	4.33
SD	1.31	1.27
Concrete senses (AvgR ≥ 4, N = 270)		
Mean	5.51	4.47
SD	0.89	1.18
Abstract senses (AvgR < 4, N = 120)		
Mean	3.24	4.03
SD	0.44	1.49

suggests that sense frequency might play a very important if not predominant role in one's perception of the concreteness of a word in general, given that the most frequent sense is listed as the first sense in WordNet. However, FirstDef has a wider and more even spread of scores across the scale, especially for abstract words. Hence words deemed concrete may more likely have a concrete first sense, while words deemed abstract may also have frequently used concrete senses. Alternatively, many abstract words (and their first senses) might be unexpectedly describable and definable like concrete concepts.

For the sense level ratings, there are 390 senses altogether, with 172 from the abstract word samples, and 218 from the concrete word samples. Table 5.6 shows the mean and standard deviation of the concreteness ratings for all senses and separately for senses deemed concrete and abstract respectively by human raters. AvgR is the average rating of the four human raters in this study, and is used as a reference here since no comparable ratings at the sense level are available from previous studies. WN refers to the scores from definitions.

Similar to word-level concreteness, the mean of human ratings on the concreteness of individual senses is higher than those obtained from definitions. Moreover, unlike the word level, human ratings for individual senses show less agreement. The variation is particularly apparent for the abstract concepts. While a dichotomous concrete-abstract distinction seems fairly feasible, the definition scores have not been able to realistically distinguish concrete or abstract concepts on a finer scale (e.g. highly concrete vs mildly concrete) as naturally as human raters. There are several factors which might account for this outcome. The range of surface structures we have attended to is quite narrow. On the one hand, the concise conventional defining styles may pose a limit on including further information with differentiae and prototypes. On the other hand, very common concrete concepts may not need so much detailed explanation as others in a dictionary. The assumption that different defining styles are more suitable for concrete concepts and abstract concepts is generally valid. For instance, we do find more definitions in the form of shunters among the abstract senses.[5] Nevertheless, how a concept is

---

[5] According to Vossen et al. (1989), "shunters" refer to where the definition for a noun is shifted from a nominal structure to a non-nominal one.

defined may also depend on how describable the concept is, which might just be one dimension amongst others contributing to the concreteness of the concept perceived by humans.

5.3.4 Current Limitations

The above results do not only echo the subjectivity in human perception of concreteness, but also suggest the multi-dimensionality of concreteness and particularly the importance of finding a systematic measure for this psychologically valid construct so that it can be operationalised and objectively studied in computational linguistics. They point to the following issues which are worth further investigation:

- The current framework for analysing the definition patterns may not have made full use of the whole range of scores, which is partly responsible for the lack of distinguishability between concrete and abstract concepts on a finer scale. For example, few definitions in our samples have both differentiae and prototype and many highly concrete meanings, mostly rated 7 by humans, may only have scored 6 from the definitions.
- In addition to making a finer distinction of the definition patterns, the subtle difference in the implicitly encoded information in the definitions should also be considered. For instance, the word "concept" (defined as "an abstract or general idea inferred or derived from specific instances") scored 6 despite the word "abstract" explicitly appears in the definition. While tangibility and image-ability may affect our judgement on concreteness, the nature of the genus could also give a hint (e.g. something defined as "a disposition...") should score less than if it is defined as "a depository...").
- Many factors including tangibility, imageability, frequency, etc., interact with one another to constitute to the perceived concreteness of a concept. It might even be too simplistic to view concreteness on a simple one-dimensional scale. For instance, attitude or cognition are apparently intangible and much more abstract than physical entities, but some concepts falling under attitude may still be more describable than others, and some concepts, though with physical existence, might be less describable than others.
- As mentioned earlier, despite the psychological validity of concreteness, studies on the mental lexicon and lexical access had at most touched on word concreteness, but seldom sense concreteness. This is particularly salient in the case of polysemy. If polysemy has been shown to affect lexical access, and concreteness plays a role in lexical storage, how do polysemy, concreteness, and lexical access impact on one another? More importantly, how does their interaction affect the way we disambiguate word senses? How should we encode such information in computational lexical resources to facilitate automatic WSD? These are all critical and practical issues for computational linguists. We shall return to them in the next chapter.

5.4 Summary

In this chapter, we have focused on a psychologically valid construct, namely concept concreteness, and explored its potential relation to the lexical sensitivity observed for WSD by analysing word association data for the relation between sense concreteness and semantic activation. The results reinforce the significance of the intrinsic nature of individual target words in WSD, and also inform the computational modelling of the mental lexicon and hence the semantic information and strategies for the task. To enable further large-scale pursuit on this property, we have tried to systematically obtain a concreteness measure from dictionary definitions. While the method needs refinement to better resemble human judgement, obviously concreteness should make a meaningful alternative classification of senses to break new ground for studying the lexical sensitivity of WSD. We will go along this line in the next chapter and propose a research agenda in this regard.

Chapter 6
Lexical Sensitivity of WSD: An Outlook

Abstract We have tried to show from our discussion in the previous chapters that while ensembles of classifiers based on supervised learning methods trained on multiple contextual features have proved to perform superiorly in current mainstream automatic word sense disambiguation, and their performance might have apparently reached a plateau, there are still considerable unknowns as far as the lexical sensitivity of the task is concerned. We have also suggested that these under-explored parts cannot be adequately addressed from the computational perspective alone, as they probably involve some intrinsic properties of words and senses, like concept concreteness, which may be cognitively based. In this final chapter, we put forth some preliminary evidence regarding the impact of concept concreteness on the information demand in disambiguation, and conclude with a research agenda which attempts to bring the two camps closer to advance the research on an area of their mutual concern.

6.1 The Unknowns of Automatic WSD

In the first three chapters, we have reviewed the development and the current status of automatic word sense disambiguation (WSD). The SENSEVAL exercises have played a key role to reveal the different issues on WSD and allow researchers to learn more about the linguistic and technical aspects of the task. We have thus been quite sure that:

- Multiple knowledge sources are required for WSD and they may have different contributions to disambiguating words of different parts-of-speech (POS).
- Systems often work better with more coarsely grained sense distinctions, and closely related senses tend to be more difficult in general.

- Supervised machine learning algorithms may differ in their performance even when trained on the same feature set, and some algorithms might be a better fit for some kinds of features.
- Given such potential bias of individual algorithms toward different features and training data, ensembles of classifiers are often used in state-of-the-art systems to overcome the weakness of single classifiers.

WSD has also long been realised as a lexically sensitive task, "in effect requiring specialized disambiguators for each polysemous word", as Resnik and Yarowsky (1997) put it. However, despite the reported performance of current systems, topping at around 70% accuracy, which is even considered reaching a plateau nowadays, we actually know very little about the lexical sensitivity of the task. Other than the facts that the number of senses and their dispersion, and perhaps also POS, would pose different levels of difficulty and associate with different degrees of effectiveness of the various knowledge sources, apparently nothing further can be said. For example:

- If some knowledge sources contribute better to target words of a particular POS, then among target words of the same POS, do the same knowledge sources contribute in a similar way?
- If POS cannot adequately differentiae the contributions of the knowledge sources, are there other ways for classifying the words and senses which may better account for the lexical sensitivity?
- If such alternative classification(s) can be identified, will their explicit modelling in WSD systems improve disambiguation performance, allowing them to overcome the current plateau?

These unanswered questions may possibly allow us to break new ground in automatic WSD. To further research on these issues, we certainly need to bring together the computational and the cognitive camps, since it is the intrinsic nature of the words and senses into which we should delve. Unlike the extrinsic factors of lexical resources and knowledge sources (as described in Chap. 1) which can be comfortably dealt with from the technical and practical perspectives, the nature of target words in terms of their individual information demand in disambiguation often involves the cognitive aspects as to how different (types of) words are organised and processed in the mental lexicon. In Chaps. 4 and 5, we have reviewed relevant psycholinguistic findings and hypothesised that concept concreteness could be a potentially valid classification of word senses bearing on the lexical sensitivity of WSD. In this chapter, we will look at more preliminary evidence to the relevance of concreteness on WSD, and propose a research agenda along this line.

6.2 Nature of Target Words

Past studies on WSD tend to pay far more attention to "positive evidence" than "neutral" or "negative" evidence. For example, a Naïve Bayes classifier using some broad semantic relations as knowledge sources for disambiguation may work fine for a word with one or more senses which have strong associations with some other words. However, not all senses of a word will significantly associate with other words. In that case, either the scores for all candidate senses are indistinguishable, or even the highest score may not be high enough for a decision to be made with confidence. Under such circumstances, the prior probabilities will become dominant and the method is inevitably reduced to one which chooses the most frequent sense. Given that it is the most frequent sense, the chance for a hit and thus the accuracy is often not bad, but such apparent effectiveness of the method could be illusory.

Although the optimal combination of knowledge sources in many studies has somehow addressed such differential information demand, a more transparent picture is still needed. It would be useful to know the relative strength of different evidence for different senses, especially for handling conflicting evidence from the context. An attempt has been made in Kwong (2000, 2005) to study how the nature of target words affect the difficulty of WSD, by analysing disambiguation results with respect to the way senses are characterised in WordNet on the one hand, and by grouping test samples according to their responses to different knowledge sources to find the commonalities within such clusters on the other.

6.2.1 Information Susceptibility

Different knowledge sources vary in their effectiveness to characterise a sense and distinguish it from other senses of the same word, and hence disambiguating it. In other words, senses have varied *information susceptibility*, which refers to the relation between the intrinsic properties of a word and the effectiveness of various types of lexico-semantic knowledge to characterise and disambiguate it. This kind of information, however, is absent from existing lexical resources. The objective of studying target nature in terms of information susceptibility is therefore to investigate the plausibility of separating words, or more precisely word senses, into fairly distinct groups (or *sense types*) according to their responses to disambiguation with different knowledge sources. The obvious problem then is "what" these different types will be. We have shown that these types would be beyond simple linguistic categories like POS, and are likely to be more semantic and perceptual. To look for such potential classifications, one way is to take a shortcut by using readily available sense classifications and see if they can account for disambiguation results in general. Alternatively, one can infer the groupings from the differential responses of senses to knowledge sources to see how senses are best characterised by different knowledge sources. The knowledge on the information susceptibility of individual target words is important for fine-tuning WSD systems and informing the optimal combination of knowledge sources for disambiguation.

6.2.2 Using Existing Semantic Classification for Sense Types

As a preliminary exploration of the possible classifications for information susceptibility relevant for WSD, Kwong (2000) tried to define sense types using the nine top nodes in the WordNet IS-A hierarchy. These top nodes are "Entity", "Psychological Feature", "Abstraction", "State", "Event", "Human Action", "Group", "Possession", and "Phenomenon". The sense type distribution and disambiguation results for various categories of texts in SEMCOR were compared, contrasting scientific writings and fiction particularly.

The most notable feature regarding sense type distribution is that "Entity" words take up only a quarter of all nouns in scientific writings, but as much as half of all nouns in fictional texts. Another frequent type is "Abstraction" (e.g. "molecular weight", "viscosity"), which occupies 30% of all noun instances in scientific writings, but only about 20% in fictional texts. Such distributional patterns, though not absolutely distinct, somehow demonstrate their dependence on text category. The analysis of the disambiguation results[1] reveals that related words in a coherent discourse are not necessarily near neighbours within a taxonomy of concepts. Words do not only relate paradigmatically (as is mostly captured in WordNet) to one another within a text. Also, assuming that fictional texts are less strongly related to particular subject areas than scientific writings, the monosemous nouns used for starting the algorithm in fictional texts may possess relatively weaker selective power toward the rest of the ambiguous words in the texts. Hence, other more reliable disambiguation information must be sought and disambiguation has to be made sensitive to the target words and the text types in question. Although defining sense types by the top nodes in the WordNet hierarchy does not show any contrasting patterns for correct and incorrect disambiguation, and therefore does not allow us to say conclusively how sense types relate to the difficulty of WSD, a significant distinction among different text categories in terms of their sense type composition can be identified, which closely correlates with the disambiguation results for different text types.

---

[1] The disambiguation was based on a narrow semantic relation measured from the taxonomic distance of two senses with respect to the WordNet hierarchy. Monosemous words were used to start a recursive filtering algorithm to gradually purge the irrelevant senses and leave only the relevant senses in a finite number of processing cycles.

6.2.3 Sense Types Clustered from Disambiguation Results

In another preliminary study, a spectrum of semantic (and other) relations with different degrees of specificity was first defined for the investigation of information susceptibility. On the specificity scale, there were six knowledge sources from the broadest to the narrowest as follows:

- *col*: simple collocation with a target sense obtained from corpus data
- *com*: overlap between the WordNet definition of a target sense and the set of words in Roget's Thesaurus heads
- *def*: overlap between the WordNet definition between two senses
- *rog*: similarity of two senses with respect to their Roget classification
- *wn*: similarity of two senses based on taxonomic distance in WordNet
- *f-2g*: flexible POS bigrams within a local window

By applying these different knowledge sources in disambiguation, the hypothesis that different groups of senses are best characterised with different types of information can be put to test.

Experimenting with a small dataset showed that most of the test samples (about 80%) were successfully disambiguated in one way or another, as compared to the most-frequent-sense baseline (53.33%). Also, most noun senses in the samples can be disambiguated by one or more types of semantic relation with other noun senses. It was found that 17 out of 60 samples could be disambiguated with the narrowest paradigmatic association. This does not necessarily mean that those cases are not susceptible to broader semantic relations, but the simple IS-A relation suffices for them. In fact, most of the samples which were correctly disambiguated with *wn* were also correctly disambiguated with one or more other information types. However, the distinction (between broad and narrow semantic relations) is important because there remains the possibility that some (untested) senses may only work with *wn*, and it is useful to gather all plausible evidence so that senses can be more confidently selected.

The results allow us to uncover more about the susceptibility of word senses to various types of information, and thus the effectiveness of such information for disambiguating them. The clusters (syntactic information only, narrow semantic information, broad semantic information, semantic information only, syntactic/semantic information, unknown) support the hypothesis that different senses are best characterised by different types of information. The clusters are unnamed, but for instance, as we go up the specificity spectrum of semantic relations (that is, from narrow to broad relations), the senses susceptible to those relations apparently involve more and more abstract thinking. Senses involving more abstract thinking tend to be disambiguated only with broader semantic relations. This observation also coincides with findings from human word association tests. For instance, in the Birkbeck word association norms (Moss and Older 1996), "loss" triggers associations like "death" and "grief", which cannot be related via a simple IS-A relation, in contrast to responses like "magic" triggered by "trick" which are simply synonymous. Such association data therefore support the claim that different strategies are employed by people to conceptualise different groups of senses, hence the idea of classifying senses by information susceptibility.

6.3 Concreteness as an Alternative Classification

Despite being a psychologically valid and intrinsic property of words and senses, concreteness is seldom addressed in WSD literature. Psychologists have shown, from lexical decision and naming tasks, that abstract words are harder to understand than concrete words, and are often acquired later. This thus implies differential underlying mechanisms in the representation, development, and processing of word meanings in the mental lexicon. By analogy, the inclusion of the concreteness information in computational lexicons should also benefit natural language processing tasks like WSD, in addition to maneuvering only linguistic and technical factors. It should also allow us to study polysemy and sense similarity in a more comprehensive and cognitively plausible way.

While there were studies investigating the relationship between lexical access and polysemy (e.g. Swinney 1979), few have addressed the relation between concreteness and polysemy, and WSD. As discussed in the last chapter, analysis on word association responses has suggested that tangible concepts seem to be more easily activated than abstract concepts; and in the case of polysemy, tangible senses appear to be more accessible than abstract senses.

It has been suggested that WSD systems should be less penalised if they fail to distinguish between closely related word senses than if they fail between distinct senses. This issue of sense similarity is addressed by Resnik and Yarowsky (1999) with quantitative characterisation in terms of sense proximity, and by Chugur et al. (2002) in terms of sense stability.

WSD is often considered a lexically sensitive task, in which individual target words might vary in their difficulty and require different treatment. Pedersen (2002) assessed the difficulty of test instances in the SENSEVAL-2 English lexical sample task by analysing the agreement among participating systems. However, with the experience from the SENSEVAL exercises, systems should have matured in one way or another to cope with the lexical sensitivity of WSD. If a difference still persists among target words within individual systems, there must be something else intrinsic to the words and senses themselves that has not been adequately recognised and effectively addressed by automatic WSD systems. We are thus interested in how lexical concreteness, as a psycholinguistic factor and an intrinsic property of words, bears on the difficulty of WSD and information demand of individual target words, in addition to other linguistic and technical factors. More importantly, we should explore how we could capitalise on such a relationship to fine-tune WSD systems and shed light on WSD evaluation.

Table 6.1 Target nouns from SENSEVAL data

SENSEVAL-1			SENSEVAL-4 (Task 17)					
Word	S	C	Word	S	C	Word	S	C
Accident	2	4.33	Area	6	4.00	Network	5	3.00
Behaviour	4	3.00	Authority	7	3.33	Order	14	2.67
Bet	2	4.50	Base	19	3.67	Part	12	3.33
Disability	1	2.00	Bill	10	5.00	People	4	6.67
Excess	4	1.50	Capital	6	4.00	Plant	4	6.67
Float	7	4.50	Carrier	11	4.67	Point	26	4.67
Giant	7	4.50	Chance	5	1.33	Policy	3	2.00
Knee	3	7.00	Condition	8	1.33	Position	16	3.00
Onion	3	6.50	Defense	11	2.33	Power	9	3.00
Promise	2	2.00	Development	9	2.67	President	4	6.00
Rabbit	3	6.50	Drug	1	6.33	Rate	4	1.33
Sack	9	4.50	Effect	6	2.00	Share	5	2.67
Scrap	4	4.00	Exchange	10	3.67	Source	9	2.33
Shirt	1	6.50	Future	3	1.33	Space	9	3.67
Steering	3	4.00	Hour	4	2.33	State	7	3.33
			Job	9	4.00	System	9	2.67
			Management	2	3.33	Value	6	1.33
			Move	5	4.00			

6.3.1 Analysing SENSEVAL Results

To explore the impact of concreteness of WSD difficulty, Kwong (2008b) compared SENSEVAL results based on the concreteness ratings of the target words. Target nouns from the English lexical sample tasks in SENSEVAL-1 and SENSEVAL-4 (Task 17) were selected and their sense definitions from WordNet 3.0 were collected. Three human judges were asked to rate the words and senses in the sample on a 7-point scale of concreteness, with 1 for highly abstract, and 7 for highly concrete. The data are listed in Table 6.1, where the column "S" refers to the number of senses in WordNet 3.0,[2] and the column "C" refers to the average of human ratings on lexical concreteness.

WSD difficulty was assumed to be reflected from system performance on individual words as shown in the task and system reports, and results summaries from SENSEVAL. For SENSEVAL-1 data, the official scores under "fine-grained, all systems, average" available from http://www.senseval.org were used. Precisions and recalls were reported, and the F1 measure was computed from them for convenience in comparison. System performance for SENSEVAL-4 (Task 17) was

---

[2] The HECTOR sense inventory was used for SENSEVAL-1, and some had very different degrees of polysemy. For example, "knee" has as many as 22 senses. On the other hand, OntoNotes senses were used for sense distinction in SENSEVAL-4 (Task 17). WordNet 3.0 senses were used as a common reference for both sets of words.

Table 6.2 Mean performance by lexical concreteness

Group	SENSEVAL-1		SENSEVAL-4 (Task 17)			
	# Samples	Mean acc	# Samples	Avg	Sys 1	Sys 4
Abstract	3	0.7235	12	77.42	85.67	75.00
Medium	8	0.6150	15	79.27	87.70	78.00
Concrete	4	0.7384	3	92.67	96.00	90.33

based on the average results on individual target words from all systems, as well as results from two individual systems (System 1 and 4, both using Support Vector Machines), reported in Pradhan et al. (2007).

6.3.2 Impact of Concreteness on WSD

Based on the average human ratings, the target words were divided into three categories along the concreteness continuum: Abstract (with average rating below 3.0), Medium (with average rating between 3.0 and 5.0 inclusive), and Concrete (with average rating above 5.0). Since only secondary data sources were used, the number of samples in each category could not be controlled, but only depended on the ratings assigned by the human judges on the target nouns. Hence the number of samples in each group is small and the distribution may not be even. Nevertheless, it happens that the datasets in SENSEVAL-1 and SENSEVAL-4 do contain examples for all three groups, allowing a preliminary analysis of the relationship between concreteness and WSD difficulty.

Table 6.2 shows the mean performance for various levels of concreteness. There were only 30 words from SENSEVAL-4 used in the comparison instead of 35, since results for five words were omitted in Pradhan et al. (2007). The mean performance with respect to the concreteness of the first sense and that based on the average concreteness of all senses were studied. The first sense was assumed to be the predominant sense according to WordNet ordering.

There are some interesting observations. From SENSEVAL-1 data, it appears that nouns at both ends of the continuum, i.e. either very concrete or very abstract, are better disambiguated than those lying in the mid-range of the continuum. With reference to the impact of the first and supposedly predominant sense or the average concreteness of all senses, a similar trend was found. Moreover, words with an abstract predominant sense or more abstract senses in general tend to be even better disambiguated than those with more concrete senses. This is an interesting phenomenon which deserves more in-depth investigation and qualitative analysis. On the other hand, analysis on SENSEVAL-4 data seems to yield results closer to our expectation. With reference to word concreteness and average concreteness of all senses, words toward the concrete side tend to be better disambiguated than those in the mid-range, which are in turn better disambiguated

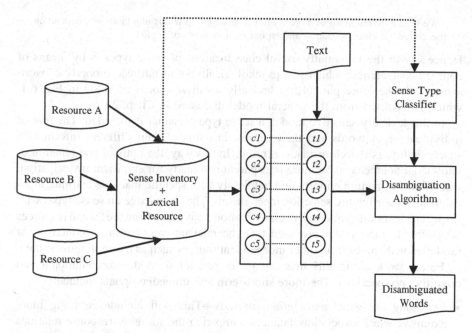

Fig. 6.1 A lexically sensitive model of WSD

than those on the abstract end. However, looking at the predominant sense, a similar situation to SENSEVAL-1 data was found, that is, those with the first sense at either end of the continuum are better disambiguated. One possible reason is that very concrete or very abstract senses are expected to occur in more characteristic linguistic contexts, which could be more successfully captured by the features used in WSD systems. Alternatively, as Pinker (2008, p. 3) suggested, "a feature of the mind … is that even our most abstract concepts are understood in terms of concrete scenarios". Hence, it remains for us to see in more qualitative terms how this concreteness effect could affect the effectiveness of various kinds of disambiguating information.

6.4 A Research Agenda

In view of the *prima facie* evidence that we have presented thus far, the lexical sensitivity of WSD can be considered a function of the information susceptibility of different types of word senses. As Wittgenstein (1958) claimed, one just needs to play the game differently with different kinds of meaning; and for classification in particular,

"We have different kinds of word ... But how we group words into kinds will depend on the aim of the classification, – and on our own inclination." (§17)

Hence one of the potentially useful classifications of sense types is by means of concept concreteness which is a psychologically valid intrinsic property of word senses. We therefore put forth a lexically sensitive model of WSD in Fig. 6.1, which is modified from the general model discussed in Chap. 2.

In this lexically sensitive model, a sense type classifier is included. The role of individual target words and their senses, in terms of their different information susceptibility, is therefore made explicit. In this way, the optimal combination of knowledge sources can be more transparent and ready for empirical investigation, instead of just resting entirely and helplessly with specific machine learning algorithms and their feature selection mechanisms. The knowledge on sense types with respect to one or other alternative classifications may come from the lexical resources accessible to the system provided that the relevant information is contained or modelled in them, or from other independent sources such as psycholinguistic data.

Future work along this line can go in parallel from the computational and cognitive perspectives. The more short-term and immediate goals include:

- *Extending the word association analysis*—This will include covering more extensive word association data and comparing the stimulus–response relations against various existing lexical resources such as semantic lexicons, ontologies, as well as large corpora, thus investigating a wider range of semantic relations and knowledge sources.
- *Obtaining a better measure for concreteness*—This would mean exploring other possibilities for operationalising the definition of sense concreteness, particularly those which are capable for distinguishing concreteness on a finer scale instead of just a dichotomous classification.
- *Gathering more psycholinguistic evidence*—More experiments on lexical access and ambiguity resolution have to be done to study the effect of lexical concreteness and sense concreteness, and simultaneously with better control on the sentential context as well as other factors including frequency, polysemy, sense relatedness, familiarity, imageability, POS, and any other intrinsic properties of words and senses bearing on the lexical sensitivity of WSD.
- *Refining contextual features for WSD*—If Schwanenflugel's (1991) context availability model holds, that is, concrete words are deemed concrete because more meaningful contexts (e.g. a sentence, a visual scene, etc.) can be readily generated by most people than for abstract words, we would expect this difference to apply to sense concreteness too, and therefore concrete senses might be associated with more distinct linguistic contexts than abstract senses. With corpus evidence, we can investigate the availability of various kinds of contextual information including collocated words and syntactic environment for senses of different degrees of concreteness, and to study how the different information contributes to their disambiguation.

With more evidence and empirical data on concept concreteness as a feasible classification of word senses in terms of their information susceptibility, the longer-term and ultimate goals would obviously be the application of such new knowledge on sense types in WSD systems to get their performance over the current plateau. More specifically, we aim at:

- *Enriching existing lexical resources*—The findings on sense concreteness and semantic activation should be incorporated into existing lexical resources to better reflect the associative distance and strength among concepts and senses, while such information will in turn contribute to better WSD.
- *Testing and fine-tuning WSD systems*—With the relevant information for the sense type classifier in our lexically sensitive model of WSD, further development and experiments can be done on automatic WSD systems to better capitalise on the target nature and more effectively handle the lexical sensitivity of the task. This will significantly address Ide and Wilks' (2006) suggestion to perfect WSD systems at least for homograph disambiguation.
- *Defining more suitable sense type classifications for WSD*—While concept concreteness is probably a potential classification relevant to the information susceptibility of word senses and thus the lexical sensitivity of WSD, there may be other relevant classifications (including POS, semantic classes, and other linguistic categories and cognitive factors) which interact with one another.

Thus the main thesis of this book is: although research on automatic WSD seems to have reached a limit, the lexical sensitivity issue is still waiting to be more fully and directly addressed. The remaining unknowns, which largely have to do with the intrinsic properties of words and senses, have to be considered from both the cognitive and computational perspectives. Our understanding on lexical sensitivity will rely on investigations into alternative classifications of word senses with respect to their information susceptibility, and such work will only be fruitful with a reunion of the two camps, not only to better inform automatic systems with more psycholinguistic evidence, but also for the mutual advancement of the two disciplines.

References

Acheson, D. J., Postle, B. R., & MacDonald, M. C. (2010). The interaction of concreteness and phonological similarity in verbal working memory. *Journal of Experimental Psychology: Learning, Memory, and Cognition, 36*(1), 17–36.

Adriaens, G., & Small, S. L. (1988). Word expert parsing revisited in a cognitive science perspective. In S. L. Small, G. W. Cottrell, & M. K. Tanenhaus (Eds.), *Lexical ambiguity resolution: Perspectives from psycholinguistics, neuropsychology, and artificial intelligence.* San Mateo: Morgan Kaufmann Publishers, Inc.

Agirre, E., & Edmonds, P. (2006). *Word sense disambiguation: Algorithms and applications.* Dordrecht: Springer.

Agirre, E., & Rigau, G. (1996). Word sense disambiguation using conceptual density. *Proceedings of the 16th International Conference on Computational Linguistics (COLING-96),* Copenhagen, Denmark (pp. 16–22).

Agirre, E., & Stevenson, M. (2006). Knowledge sources for WSD. In E. Agirre & P. Edmonds (Eds.), *Word sense disambiguation: Algorithms and applications.* Dordrecht: Springer.

Aitchison, J. (2003). *Words in the mind: An introduction to the mental lexicon.* Malden: Blackwell Publishing Ltd.

Amsler, R. (1981). A taxonomy for English nouns and verbs. *Proceedings of the 19th Annual Meeting of the Association for Computational Linguistics (ACL '81),* Stanford (pp. 133–138).

Atkins, B. T. S., & Rundell, M. (2008). *The Oxford guide to practical lexicography.* Oxford: Oxford University Press.

Bleasdale, F. A. (1987). Concreteness dependent associative priming: Separate lexical organization for concrete and abstract words. *Journal of Experimental Psychology: Learning, Memory, and Cognition, 13,* 582–594.

Bock, K., & Levelt, W. (1994). Language production: Grammatical encoding. In M. A. Gernsbacher (Ed.), *Handbook of psycholinguistics.* San Diego: Academic Press.

Brown, P. F., Della Pietra, S. A., Della Pietra, V. J., & Mercer, R. L. (1991). Word-sense disambiguation using statistical methods. *Proceedings of the 29th Annual Meeting of the Association for Computational Linguistics (ACL '91),* Berkeley, CA (pp. 264–270).

Brown, S. W. (2008). Choosing sense distinctions for WSD: Psycholinguistic evidence. *Proceedings of the 46th Annual Meeting of the Association for Computational Linguistics: Human Language Technologies (ACL-08:HLT), Short Papers (Companion Volume),* Columbus, Ohio, USA (pp. 249–252).

Bruce, R., & Guthrie, L. (1992). Genus disambiguation: A study in weighted preference. *Proceedings of the 14th International Conference on Computational Linguistics (COLING '92),* Nantes, France (pp. 1187–1191).

Bruce, R., & Wiebe, J. (1994). Word-sense disambiguation using decomposable models. *Proceedings of the 32nd Annual Meeting of the Association for Computational Linguistics (ACL '94)*, Las Cruces, New Mexico (pp. 139–145).

Calzolari, N. (1988). The dictionary and the thesaurus can be combined. In M. Evens (Ed.), *Relational models of the lexicon*. Cambridge: Cambridge University Press.

Caramazza, A. (1997). How many levels of processing are there in lexical access? *Cognitive Neuropsychology, 14*, 177–208.

Caraballo, S. A. (1999). Automatic construction of a hypernym-labeled noun hierarchy. *Proceedings of the 37th Annual Meeting of the Association for Computational Linguistics (ACL '99)*, College Park, Maryland (pp. 120–126).

Carroll, D. W. (2004). *Psychology of language*. Belmont, CA: Thomson Wadsworth.

Chan, Y. S., Ng, H. T., & Chiang, D. (2007). Word sense disambiguation improves statistical machine translation. *Proceedings of the 45th Annual Meeting of the Association for Computational Linguistics (ACL 2007)*, Prague, Czech Republic (pp. 33–40).

Chen, J. N., & Chang, J. S. (1998). Topical clustering of MRD senses based on information retrieval techniques. *Computational Linguistics, 24*(1), 61–95.

Chklovski, T., & Mihalcea, R. (2002). Building a sense tagged corpus with Open Mind Word Expert. *Proceedings of the Workshop on Word Sense Disambiguation: Recent Successes and Future Directions*, Philadelphia (pp. 116–123).

Chodorow, M. S., Byrd, R. J., & Heidorn, G. E. (1985). Extracting semantic hierarchies from a large on-line dictionary. *Proceedings of the 23rd Annual Meeting of the Association for Computational Linguistics (ACL '85)*, Chicago (pp. 299–304).

Chugur, I., Gonzalo, J., & Verdejo, F. (2002). Polysemy and sense proximity in the Senseval-2 test suite. *Proceedings of the Workshop on Word Sense Disambiguation: Recent Successes and Future Directions*, Philadelphia (pp. 32–39).

Church, K. W., & Hanks, P. (1990). Word association norms, mutual information, and lexicography. *Computational Linguistics, 16*(1), 22–29.

Church, K. W., Gale, W., Hanks, P., Hindle, D., & Moon, R. (1994). Lexical substitutability. In B. T. S. Atkins & A. Zampolli (Eds.), *Computational approaches to the lexicon*. Oxford: Oxford University Press.

Ciaramita, M., & Johnson, M. (2003). Supersense tagging of unknown nouns in WordNet. *Proceedings of the 2003 Conference on Empirical Methods in Natural Language Processing*, Sapporo, Japan.

Clark, E. V., & Clark, H. H. (1979). When nouns surface as verbs. *Language, 55*(4), 767–811.

Collins, A. M., & Loftus, E. F. (1975). A spreading-activation theory of semantic processing. *Psychological Review, 82*(6), 407–428.

Collins, A. M., & Quillian, M. R. (1969). Retrieval time from semantic memory. *Journal of Verbal Learning and Verbal Behavior, 8*, 240–247.

Collins, A. M., & Quillian, M. R. (1970). Does category size affect categorization time? *Journal of Verbal Learning and Verbal Behavior, 9*, 432–438.

Copestake, A., & Briscoe, E. J. (1995). Semi-productive polysemy and sense extension. *Journal of Semantics, 12*, 15–67.

Covington, M. A. (1994). *Natural Language Processing for Prolog Programmers*. Englewood Cliffs: Prentice Hall.

Cowie, J., Guthrie, J., & Guthrie, L. (1992). Lexical disambiguation using simulated annealing. *Proceedings of the 14th International Conference on Computational Linguistics (COLING-92)*, Nantes, France (pp. 359–365).

Cruse, D.A. (1986). *Lexical semantics*. Cambridge: Cambridge University Press.

Curran, J.R. (2005). Supersense tagging of unknown nouns using semantic similarity. *Proceedings of the 43rd Annual Meeting of the Association for Computational Linguistics (ACL 2005)*, Ann Arbor, Michigan.

Dagan, I., & Itai, A. (1994). Word sense disambiguation using a second language monolingual corpus. *Computational Linguistics, 20*(4), 563–596.

Dagan, I., Lee, L., & Pereira, F. (1997). Similarity-based methods for word sense disambiguation. *Proceedings of the 35th Annual Meeting of the Association for Computational Linguistics and 8th Conference of the European Chapter of the Association for Computational Linguistics (ACL/EACL '97)*, Madrid, Spain (pp. 56–63).

Dang, H. T., & Palmer, M. (2005). The role of semantic roles in disambiguating verb senses. *Proceedings of the 43rd Annual Meeting of the Association for Computational Linguistics (ACL 2005)*, Ann Arbor, Michigan (pp. 42–49).

de Groot, A. M. B. (1989). Representational aspects of word imageability and word frequency as assessed through word association. *Journal of Experimental Psychology: Learning, Memory, and Cognition, 15*(5), 824–845.

Diab, M. (2004). Relieving the data acquisition bottleneck in word sense disambiguation. *Proceedings of the 42nd Annual Meeting of the Association for Computational Linguistics (ACL 2004)*, Barcelona, Spain.

Diab, M., & Resnik, P. (2002). An unsupervised method for word sense tagging using parallel corpora. *Proceedings of the 40th Annual Meeting of the Association for Computational Linguistics (ACL 2002)*, Philadelphia (pp. 255–262).

Edmonds, P., & Cotton, S. (2001). SENSEVAL-2: Overview. *Proceedings of the 2nd International Workshop on Evaluating Word Sense Disambiguation Systems (SENSEVAL-2)*, Toulouse, France (pp. 1–6).

Erk, K., McCarthy, D., & Gaylord, N. (2009). Investigations on word senses and word usages. *Proceedings of the 47th Annual Meeting of the Association for Computational Linguistics and the 4th International Joint Conference on Natural Language Processing (ACL-IJCNLP 2009)*, Singapore (pp. 10–18).

Evens, M. W. (1988). *Relational models of the lexicon: Representing knowledge in semantic networks*. Cambridge: Cambridge University Press.

Fellbaum, C. (1998). *WordNet: An electronic lexical database*. Cambridge: The MIT Press.

Florian, R., Cucerzan, S., Schafer, C., & Yarowsky, D. (2002). Combining classifiers for word sense disambiguation. *Natural Language Engineering, 8*(4), 327–341.

Fontenelle, T. (1997). Using a bilingual dictionary to create semantic networks. *International Journal of Lexicography, 10*(4), 275–303.

Foss, D. J. (1970). Some effects of ambiguity upon sentence comprehension. *Journal of Verbal Learning and Verbal Behavior, 9*, 699–706.

Gale, W., Church, K. W., & Yarowsky, D. (1992a). One sense per discourse. *Proceedings of the Speech and Natural Language Workshop*, San Francisco, CA (pp. 233–237).

Gale, W., Church, K. W., & Yarowsky, D. (1992b). Estimating upper and lower bounds on the performance of word-sense disambiguation programs. *Proceedings of the 30th Annual Meeting of the Association for Computational Linguistics (ACL '92)*, University of Delaware, Newark, DE (pp. 249–256).

Galton, F. (1883). *Inquiries into human faculty and its development*. London: Dent.

Gliozzo, A., Giuliano, C., & Strapparava, C. (2005). Domain kernels for word sense disambiguation. *Proceedings of the 43rd Annual Meeting of the Association for Computational Linguistics (ACL 2005)*, Ann Arbor, Michigan (pp. 403–410).

Grozea, C. (2004). Finding optimal parameter settings for high performance word sense disambiguation. *Proceedings of the 3rd International Workshop on the Evaluation of Systems for the Semantic Analysis of Text (SENSEVAL-3)*, Barcelona, Spain (pp. 125–128).

Guthrie, L., Slator, B.M., Wilks, Y., & Bruce, R. (1990). Is there content in empty heads? *Proceedings of the 13th International Conference on Computational Linguistics (COLING-90)*, Helsinki, Finland (pp. 138–143).

Hanks, P. (1987). Definitions and explanations. In J. M. Sinclair (Ed.), *Looking up: An account of the COBUILD project in lexical computing*. London: HarperCollins Publishers.

Hanks, P. (2000). Do word meanings exist? *Computers and the Humanities, 34*(1–2), 205–215.

Harley, A., & Glennon, D. (1997). Sense tagging in action: Combining different tests with additive weightings. *Proceedings of SIGLEX '97 Workshop: Tagging Text with Lexical Semantics: Why, What, and How?* Washington, DC (pp. 74–78).

Hayes, P. J. (1977). On semantic nets, frames and associations. *Proceedings of the 5th International Joint Conference on Artificial Intelligence (IJCAI-77)*, Cambridge, MA (pp. 99–107).

Hirsh, K. W., & Tree, J. J. (2001). Word association norms for two cohorts of British adults. *Journal of Neurolinguistics, 14*, 1–44.

Hirst, G. (1987). *Semantic interpretation and the resolution of ambiguity.* Cambridge: Cambridge University Press.

Hurford, J. R., & Heasley, B. (1983). *Semantics: A coursebook.* Cambridge: Cambridge University Press.

Ide, N., & Fellbaum, C. (2006). Introduction. *Proceedings of the Workshop on Making Sense of Sense: Bringing Psycholinguistics and Computational Linguistics Together*, Trento, Italy.

Ide, N., & Veronis, J. (1998). Introduction to the special issue on word sense disambiguation: The state of the art. *Computational Linguistics, 24*(1), 1–40.

Ide, N., & Wilks, Y. (2006). Making sense about sense. In E. Agirre & P. Edmonds (Eds.), *Word sense disambiguation: Algorithms and applications.* Dordrecht: Springer.

Jefferies, E., Patterson, K., Jones, R. W., & Lambon Ralph, M. A. (2009). Comprehension of concrete and abstract words in semantic dementia. *Neuropsychology, 23*(4), 492–499.

Jenkins, J. J. (1970). The 1952 Minnesota word association norms. In L. Postman & G. Keppel (Eds.), *Norms of word association.* New York: Academic Press.

Johansson, R., & Nugues, P. (2008). Dependency-based syntactic-semantic analysis with PropBank and NomBank. *Proceedings of the 12th Conference on Computational Natural Language Learning (CoNLL 2008)*, Manchester (pp. 183–187).

Jorgensen, J. (1990). The psychological reality of word senses. *Journal of Psycholinguistic Research, 19*, 167–190.

Joyce, T. (2005). Constructing a large-scale database of Japanese word associations. (Special issue: Kanji corpus research, edited by Katsuo Tamaoka), *Glottometrics, 10.*

Karov, Y., & Edelman, S. (1998). Similarity-based word sense disambiguation. *Computational Linguistics, 24*(1), 41–59.

Kilgarriff, A. (1992). *Polysemy.* PhD Dissertation. University of Sussex.

Kilgarriff, A. (1993). Dictionary word sense distinctions: An enquiry into their nature. *Computers and the Humanities, 26*(5–6), 365–387.

Kilgarriff, A. (1997). "I don't believe in word senses". *Computers and the Humanities, 31*(2), 91–113.

Kilgarriff, A. (1998). Gold standard datasets for evaluating word sense disambiguation programs. *Computer Speech and Language, Special Issue on Evaluation, 12*(3), 453–472.

Kilgarriff, A. (1999). 95% replicability for manual word sense tagging. *Proceedings of the 9th Conference of the European Chapter of the Association for Computational Linguistics (EACL '99)*, Bergen, Norway (pp. 277–278).

Kilgarriff, A. (2006). Word senses. In E. Agirre & P. Edmonds (Eds.), *Word sense disambiguation: Algorithms and applications.* Dordrecht: Springer.

Kilgarriff, A., & Palmer, M. (2000). Introduction to the special issue on Senseval. *Computers and the Humanities, 34*(1–2), 1–13.

Kilgarriff, A., & Rosenzweig, J. (1999). English SENSEVAL: Reports and results. *Proceedings of the 5th Natural Language Processing Pacific Rim Symposium (NLPRS '99)*, Beijing, China.

Klavans, J., Chodorow, M., & Wacholder, N. (1990). From dictionary to knowledge base via taxonomy. *Proceedings of the 6th Conference of the University of Waterloo*, Canada (pp. 110–132).

Klein, D. E., & Murphy, G. L. (2001). The representation of polysemous words. *Journal of Memory and Language, 45*, 259–282.

Klein, D. E., & Murphy, G. L. (2002). Paper has been my ruin: Conceptual relations of polysemous senses. *Journal of Memory and Language, 47*, 548–570.

Klepousniotou, E. (2002). The processing of lexical ambiguity: Homonymy and polysemy in the mental lexicon. *Brain and Language, 81*, 205–223.

Knight, K., & Luk, S.K. (1994). Building a large-scale knowledge base for machine translation. *Proceedings of the Twelfth National Conference on Artificial Intelligence (AAAI-94)*, Seattle, Washington, DC (pp. 773–778).

Kohomban, U. S., & Lee, W. S. (2005). Learning semantic classes for word sense disambiguation. *Proceedings of the 43rd Annual Meeting of the Association for Computational Linguistics (ACL 2005)*, Ann Arbor, Michigan (pp. 34–41).

Korhonen, A., & Preiss, J. (2003). Improving subcategorization acquisition using word sense disambiguation. *Proceedings of the 41st Annual Meeting of the Association for Computational Linguistics (ACL 2003)*, Sapporo, Japan.

Kousta, S.-T., Vigliocco, G., Vinson, D. P., Andrews, M., & Del Campo, E. (2011). The representation of abstract words: Why emotion matters. *Journal of Experimental Psychology: General, 140*(1), 14–34.

Kroll, J. F., & Merves, J. S. (1986). Lexical access for concrete and abstract words. *Journal of Experimental Psychology: Learning, Memory, and Cognition, 12*, 92–107.

Krovetz, R., & Croft, W. B. (1992). Lexical ambiguity and information retrieval. *ACM Transactions on Information Systems, 10*(2), 115–141.

Kucera, H., & Francis, W. N. (1967). *Computational analysis of present-day American English*. Providence: Brown University Press.

Kwong, O. Y. (1998). Bridging the gap between dictionary and thesaurus. *Proceedings of the 36th Annual Meeting of the Association for Computational Linguistics and 17th International Conference on Computational Linguistics (COLING-ACL '98)*, Montréal, Canada.

Kwong, O. Y. (2000). Word sense selection in texts: An integrated model. *Technical Report UCAM-CL-TR-504*. Computer Laboratory, University of Cambridge.

Kwong, O. Y. (2005). Word sense classification based on information susceptibility. In A. Lenci, S. Montemagni, & V. Pirrelli (Eds.), *Acquisition and representation of word meaning*, Linguistica Computazionale (pp. 89–115).

Kwong, O. Y. (2007). Sense abstractness, semantic activation and word sense disambiguation: Implications from word association norms. *Proceedings of the 4th International Workshop on Natural Language Processing and Cognitive Science (NLPCS 2007)*, Funchal, Madeira, Portugal (pp. 169–178).

Kwong, O. Y. (2008a). A preliminary study on inducing lexical concreteness from dictionary definitions. *Proceedings of the 5th International Workshop on Natural Language Processing and Cognitive Science (NLPCS 2008)*, Barcelona, Spain (pp. 84–93).

Kwong, O. Y. (2008b). A preliminary study on the impact of lexical concreteness on word sense disambiguation. *Proceedings of the 22nd Pacific Asia Conference on Language, Information and Computation (PACLIC 22)*, Cebu, Philippines (pp. 235–244).

Kwong, O. Y. (2009). Sense abstractness, semantic activation and word sense disambiguation. *International Journal of Speech Technology, 11*, 135–146.

Kwong, O. Y. (2011). Measuring concept concreteness from the lexicographic perspective. *Proceedings of the 25th Pacific Asia Conference on Language, Information and Computation (PACLIC 25)*, Singapore.

Landes, S., Leacock, C., & Tengi, R. I. (1998). Building semantic concordances. In C. Fellbaum (Ed.), *WordNet: An electronic lexical database*. Cambridge: The MIT Press.

Lapata, M., & Brew, C. (2004). Verb class disambiguation using informative priors. *Computational Linguistics, 30*(1), 45–73.

Leacock, C., Chodorow, M., & Miller, M. A. (1998). Using corpus statistics and WordNet relations for sense identification. *Computational Linguistics, 24*(1), 147–165.

Leacock, C., Towell, G., & Voorhees, E. (1993). Corpus-based statistical sense resolution. *Proceedings of the ARPA Human Language Technology Workshop*, Princeton, NJ (pp. 260–265).

Lee, Y. K., & Ng, H. T. (2002). An empirical evaluation of knowledge sources and learning algorithms for word sense disambiguation. *Proceedings of the 2002 Conference on Empirical Methods in Natural Language Processing (EMNLP 2002)*, Philadelphia (pp. 41–48).

Lee, Y. K., Ng, H. T., & Chia, T. K. (2004). Supervised word sense disambiguation with support vector machines and multiple knowledge sources. *Proceedings of the 3rd International Workshop on the Evaluation of Systems for the Semantic Analysis of Text (SENSEVAL-3)*, Barcelona, Spain.

Lesk, M. E. (1986). Automatic sense disambiguation using machine readable dictionaries: How to tell a pine cone from an ice cream cone. *Proceedings of the 1986 SIGDOC Conference*, Toronto, Canada (pp. 24–26).

Levin, B. (1993). *English verb classes and alternations: A preliminary investigation*. Chicago: University of Chicago Press.

Lin, D. (1997). Using syntactic dependency as local context to resolve word sense ambiguity. *Proceedings of the 35th Annual Meeting of the Association for Computational Linguistics and 8th Conference of the European Chapter of the Association for Computational Linguistics (ACL/EACL '97)*, Madrid, Spain (pp. 64–71).

Lin, D. (1998). Automatic retrieval and clustering of similar words. *Proceedings of the 36th Annual Meeting of the Association for Computational Linguistics and 17th International Conference on Computational Linguistics (COLING-ACL '98)*, Montréal, Canada.

Lyons, J. (1981). *Language and linguistics: An introduction*. Cambridge: Cambridge University Press.

Lytinen, S. L. (1988). Are vague words ambiguous? In S. L. Small, G. W. Cottrell, & M. K. Tanenhaus (Eds.), *Lexical ambiguity resolution: Perspectives from psycholinguistics, neuropsychology, and artificial intelligence*. San Mateo: Morgan Kaufmann Publishers, Inc.

Magnini, B., Strapparava, C., Pezzulo, G., & Gliozzo, A. (2002). The role of domain information in word sense disambiguation. *Natural Language Engineering, 8*(4), 359–373.

Markowitz, J., Ahlswede, T., & Evens, M. (1986). Semantically significant patterns in dictionary definitions. *Proceedings of the 24th Annual Meeting of the Association for Computational Linguistics (ACL '86)*, New York (pp. 112–119).

Màrquez, L., Escudero, G., Martínez, D., & Rigau, G. (2006). Supervised corpus-based methods for WSD. In E. Agirre & P. Edmonds (Eds.), *Word sense disambiguation: Algorithms and applications*. Dordrecht: Springer.

Martínez, D., Agirre, E., & Màrquez, L. (2002). Syntactic features for high precision word sense disambiguation. *Proceedings of the 19th International Conference on Computational Linguistics (COLING 2002)*, Taipei, Taiwan.

McCarthy, D. (1997). Word sense disambiguation for acquisition of selectional preferences. *Proceedings of the Workshop on Automatic Information Extraction and Building Lexical Semantic Resources for NLP Applications*, Madrid, Spain (pp. 52–60).

McCarthy, D. (2006). Relating WordNet senses for word sense disambiguation. *Proceedings of the EACL-2006 Workshop on Making Sense of Sense: Bringing Psycholinguistics and Computational Linguistics Together*, Trento, Italy (pp. 17–24).

McCarthy, D., & Carroll, J. (2003). Disambiguating nouns, verbs, and adjectives using automatically acquired selectional preferences. *Computational Linguistics, 29*(4), 639–654.

McCarthy, D., Koeling, R., Weeds, J., & Carroll, J. (2007). Unsupervised acquisition of predominant word senses. *Computational Linguistics, 33*(4), 553–590.

McClelland, J. L., & Rumelhart, D. E. (1981). An interactive activation model of context effects in letter perception: Part 1. An account of basic findings. *Psychological Review, 88*, 375–407.

McHale, M. L., & Crowter, J. J. (1994). Constructing a lexicon from a machine readable dictionary. *Technical Report RL-TR-94-178*, Rome Laboratory, Griffiss Air Force Base, New York.

McRoy, S. W. (1992). Using multiple knowledge sources for word sense disambiguation. *Computational Linguistics, 18*(1), 1–30.

Melinger, A., Schulte im Walde, S., & Weber, A. (2006). Characterizing response types and revealing noun ambiguity in German association norms. *Proceedings of the EACL-2006 Workshop on Making Sense of Sense: Bringing Psycholinguistics and Computational Linguistics Together*, Trento, Italy (pp. 41–48).

Mihalcea, R. F. (2002). Word sense disambiguation with pattern learning and automatic feature selection. *Natural Language Engineering, 8*(4), 343–358.

Mihalcea, R. (2006). Knowledged-based methods for WSD. In E. Agirre & P. Edmonds (Eds.), *Word sense disambiguation: Algorithms and applications*. Dordrecht: Springer.

Mihalcea, R., & Moldovan, D. I. (1999a). A method for word sense disambiguation of unrestricted text. *Proceedings of the 37th Annual Meeting of the Association for Computational Linguistics (ACL'99)*, College Park, Maryland (pp. 152–158).

Mihalcea, R., & Moldovan, D. I. (1999b). An automatic method for generating sense tagged corpora. *Proceedings of the 16th National Conference on Artificial Intelligence (AAAI-99)*, Orlando, Florida (pp. 461–466).

Mihalcea, R., & Pedersen, T. (2005). *Advances in word sense disambiguation*. Ann Arbor: Tutorial at ACL 2005.

Mihalcea, R., Chklovski, T., & Kilgarriff, A. (2004). The SENSEVAL-3 English lexical sample task. *Proceedings of the 3rd International Workshop on the Evaluation of Systems for the Semantic Analysis of Text (SENSEVAL-3)*, Barcelona, Spain (pp. 25–28).

Miller, G. A., & Charles, W. G. (1991). Contextual correlates of semantic similarity. *Language and Cognitive Processes, 6*(1), 1–28.

Miller, G. A., Beckwith, R., Fellbaum, C., Gross, D., & Miller, K. J. (1990). Introduction to WordNet: An on-line lexical database. *International Journal of Lexicography, 3*(4), 235–244.

Morton, J. (1969). Interaction of information in word recognition. *Psychological Review, 76*, 165–178.

Moss, H., & Older, L. (1996). *Birkbeck word association norms*. Hove: Psychology Press.

Nakamura, J., & Nagao, M. (1988). Extraction of semantic information from an ordinary English dictionary and its evaluation. *Proceedings of the 12th International Conference on Computational Linguistics (COLING '88)*, Budapest, Hungary (pp. 459–464).

Navigli, R. (2006a). Consistent validation of manual and automatic sense annotations with the aid of semantic graphs. *Computational Linguistics, 32*(2), 273–281.

Navigli, R. (2006b). Meaningful clustering of senses helps boost word sense disambiguation performance. *Proceedings of the 21st International Conference on Computational Linguistics and 44th Annual Meeting of the Association for Computational Linguistics (COLING-ACL 2006)*, Sydney (pp. 105–112).

Navigli, R. (2009). Word sense disambiguation: A survey. *ACM Computing Surveys, 41*(2), 1–69.

Navigli, R., & Crisafulli, G. (2010). Inducing word senses to improve web search result clustering. *Proceedings of the 2010 Conference on Empirical Methods in Natural Language Processing*, MIT, Massachusetts, USA (pp. 116–126).

Ng, H. T. (1997). Getting serious about word sense disambiguation. *Proceedings of SIGLEX '97 Workshop: Tagging Text with Lexical Semantics: Why, What, and How?*, Washington, DC.

Ng, H. T., & Lee, H. B. (1996). Integrating multiple knowledge sources to disambiguate word senses: An exemplar-based approach. *Proceedings of the 34th Annual Meeting of the Association for Computational Linguistics (ACL'96)*, Santa Cruz, CA (pp. 40–47).

Ng, H. T., Wang, B., & Chan, Y. S. (2003). Exploiting parallel texts for word sense disambiguation: An empirical study. *Proceedings of the 41st Annual Meeting of the Association for Computational Linguistics (ACL 2003)*, Sapporo, Japan (pp. 455–462).

Niles, I., & Pease, A. (2001). Towards a standard upper ontology. *Proceedings of the 2nd International Conference on Formal Ontology in Information Systems (FOIS-2001)*, Ogunquit, Maine.

Onifer, W., & Swinney, D. A. (1981). Accessing lexical ambiguities during sentence comprehension: Effects of frequency of meaning and contextual bias. *Memory and Cognition, 9*, 225–236.

Paivio, A. (1986). *Mental representation: A dual coding approach*. Oxford: Oxford University Press.

Paivio, A., Yuille, J. C., & Madigan, S. A. (1968). Concreteness, imagery, and meaningfulness values for 925 nouns. *Journal of Experiment Psychology, Monograph Supplement, 76*(1, Pt.2), 1–25.

Palmer, M., Ng, H. T., & Dang, H. T. (2006). Evaluation of WSD systems. In E. Agirre & P. Edmonds (Eds.), *Word sense disambiguation: Algorithms and applications*. Dordrecht: Springer.

Pease, A., Niles, I., & Li, J. (2002). The Suggested Upper Merged Ontology: A large ontology for the semantic web and its applications. *Working Notes of the AAAI-2002 Workshop on Ontologies and the Semantic Web*, Edmonton, Canada.

Pedersen, T. (2000). A simple approach to building ensembles of naive Bayesian classifiers for word sense disambiguation. *Proceedings of the 1st Meeting of the North American Chapter of the Association for Computational Linguistics (NAACL 2000)*, Seattle, Washington, DC (pp. 63–69).

Pedersen, T. (2002). Assessing system agreement and instance difficulty in the lexical sample tasks of SENSEVAL-2. *Proceedings of the Workshop on Word Sense Disambiguation: Recent Successes and Future Directions*, Philadelphia, PA, USA (pp. 40–46).

Pedersen, T. (2006). Unsupervised corpus-based methods for WSD. In E. Agirre & P. Edmonds (Eds.), *Word sense disambiguation: Algorithms and Applications*. Dordrecht: Springer.

Pedersen, T., & Bruce, R. (1998). Knowledge lean word sense disambiguation. *Proceedings of the 15th National Conference on Artificial Intelligence (AAAI-98)*, Madison, USA (pp. 800–805).

Pinker, S. (2008). *The stuff of thought*. London: Penguin Books Ltd.

Plaza, L., Jimeno-Yepes, A. J., Díaz, A., & Aronson, A. R. (2011). Studying the correlation between different word sense disambiguation methods and summarization effectiveness in biomedical texts. *BMC Bioinformatics, 12*, 355.

Ponzetto, S. P., & Navigli, R. (2010). Knowledge-rich word sense disambiguation rivaling supervised systems. *Proceedings of the 48th Annual Meeting of the Association for Computational Linguistics (ACL 2010)*, Uppsala, Sweden (pp. 1522–1531).

Pradhan, S. S., Loper, E., Dligach, D., & Palmer, M. (2007). SemEval-2007 Task 17: English lexical sample, SRL and all words. *Proceedings of the 4th International Workshop on Semantic Evaluations (SemEval-2007)*, Prague, Czech Republic (pp. 87–92).

Procter, P. (1978). *Longman dictionary of contemporary English*. Longman Group Ltd.

Pustejovsky, J. (1991). The generative lexicon. *Computational Linguistics, 17*(4), 409–441.

Pustejovsky, J., & Boguraev, B. (1993). Lexical knowledge representation and natural language processing. *Artificial Intelligence, 63*, 193–223.

Quillian, M. R. (1968). Semantic memory. In M. Minsky (Ed.), *Semantic information processing*. Cambridge, MA: MIT Press.

Rada, R., Mili, H., Bicknell, E., & Blettner, M. (1989). Development and application of a metric on semantic nets. *IEEE Transactions on Systems, Man, and Cybernetics, 19*(1), 17–30.

Rayner, K., & Frazier, L. (1989). Selection mechanisms in reading lexically ambiguous words. *Journal of Experimental Psychology: Learning, Memory, and Cognition, 15*(5), 779–790.

Resnik, P. (1993). *Selection and information: A class-based approach to lexical relationships*. Doctoral Dissertation, Department of Computer and Information Science, University of Pennsylvania.

Resnik, P. (1995a). Using information content to evaluate semantic similarity in a taxonomy. *Proceedings of the 14th International Joint Conference on Artificial Intelligence (IJCAI-95)*, Montréal, Canada (pp. 448–453).

Resnik, P. (1995b). Disambiguating noun groupings with respect to WordNet senses. *Proceedings of the Third Workshop on Very Large Corpora*, Massachusetts, USA (pp. 54–68).

Resnik, P. (1997). Selection preference and sense disambiguation. *Proceedings of SIGLEX '97 Workshop: Tagging Text with Lexical Semantics: Why, What, and How?*, Washington, DC.

Resnik, P. (2006). WSD in NLP applications. In E. Agirre & P. Edmonds (Eds.), *Word sense disambiguation: Algorithms and application*. Dordrecht: Springer.

Resnik, P., & Yarowsky, D. (1997). A perspective on word sense disambiguation methods and their evaluation. *Proceedings of SIGLEX '97 Workshop: Tagging Text with Lexical Semantics: Why, What, and How?*, Washington, DC (pp. 79–86).

Resnik, P., & Yarowsky, D. (1999). Distinguishing systems and distinguishing senses: New evaluation methods for word sense disambiguation. *Natural Language Engineering, 5*(2), 113–133.

Rigau, G., Atserias, J., & Agirre, E. (1997). Combining unsupervised lexical knowledge methods for word sense disambiguation. *Proceedings of the 35th Annual Meeting of the Association for Computational Linguistics and the 8th Conference of the European Chapter of the Association for Computational Linguistics (ACL/EACL '97)*, Madrid, Spain (pp. 48–55).

Riloff, E., & Jones, R. (1999). Learning dictionaries for information extraction by multi-level bootstrapping. *Proceedings of the Sixteenth National Conference on Artificial Intelligence (AAAI-99)*, Orlando, Florida (pp. 474–479).

Riloff, E., & Shepherd, J. (1999). A corpus-based bootstrapping algorithm for semi-automated semantic lexicon construction. *Natural Language Engineering, 5*(2), 147–156.

Roark, B., & Charniak, E. (1998). Noun-phrase co-occurrence statistics for semi-automatic semantic lexicon construction. *Proceedings of the 36th Annual Meeting of the Association for Computational Linguistics and 17th International Conference on Computational Linguistics (COLING-ACL '98)*, Montréal, Canada (pp. 1110–1116).

Rodd, J., Gaskell, G., & Marslen-Wilson, W. (2002). Making sense of semantic ambiguity: Semantic competition in lexical access. *Journal of Memory and Language, 46*, 245–266.

Roth, M., & Schulte im Walde, S. (2008). Corpus co-occurrence, dictionary and wikipedia entries as resources for semantic relatedness information. *Proceedings of the 6th Conference on Language Resources and Evaluation*, Marrakech, Morocco.

Sanderson, M. (1994). Word sense disambiguation and information retrieval. *Proceedings of the 17th Annual International ACM SIGIR Conference on Research and Development in Information Retrieval*, Dublin, Ireland (pp. 142–151).

Schulte im Walde, S., Melinger, A., Roth, M., & Weber, A. (2008). An empirical characterisation of response types in German association norms. *Research on Language and Computation, 6*(2), 205–238.

Schütze, H. (1992). Dimensions of meaning. *Proceedings of Supercomputing*, Minneapolis, MN (pp. 787–796).

Schütze, H. (1998). Automatic word sense discrimination. *Computational Linguistics, 24*(1), 97–123.

Schwanenflugel, P. J. (1991). Why are abstract concepts hard to understand? In P. J. Schwanenflugel (Ed.), *The psychology of word meanings*. Hillsdale, NJ: Lawrence Erlbaum Associates, Inc.

Seidenberg, M. S., Tanenhaus, M. K., Leiman, J. M., & Bienkowski, M. (1982). Automatic access of the meanings of ambiguous words in context: Some limitations of knowledge-based processing. *Cognitive Psychology, 14*, 489–537.

Simpson, G. B., & Burgess, C. (1985). Activation and selection processes in the recognition of ambiguous words. *Journal of Experimental Psychology: Human Perception and Performance, 11*, 28–39.

Sinclair, J. (1987). *Collins COBUILD English language dictionary*. London: HarperCollins.

Small, S., & Rieger, C. (1982). Parsing and comprehending with word experts (A theory and its realization). In W. G. Lehnert & M. H. Ringle (Eds.), *Strategies for natural language processing*. New Jersey: Lawrence Erlbaum Associates.

Snyder, B., & Palmer, M. (2004). The English all-words task. *Proceedings of the 3rd International Workshop on the Evaluation of Systems for the Semantic Analysis of Text (SENSEVAL-3)*, Barcelona, Spain (pp. 41–43).

Sparck Jones, K. (1986). *Synonymy and semantic classification*. Edinburgh: Edinburgh University Press.

Specia, L., das Graças Volpe Nunes, M., Stevenson, M., & Ribeiro, G. C. B. (2006). Multilingual versus monolingual WSD. *Proceedings of the EACL-2006 Workshop on Making Sense of Sense: Bringing Psycholinguistics and Computational Linguistics Together*, Trento, Italy (pp. 33–40).

Specia, L., Stevenson, M., & das Graças Volpe Nunes, M. (2010). Assessing the contribution of shallow and deep knowledge sources for word sense disambiguation. *Language Resources and Evaluation, 44*(4), 295–313.

Stevenson, M. (2003). *Word sense disambiguation: The case for combinations of knowledge sources*. Stanford, California: CSLI Publications.

Stevenson, M., & Wilks, Y. (1999). Combining weak knowledge sources for sense disambiguation. *Proceedings of the 16th International Joint Conference on Artificial Intelligence (IJCAI-99)*, Stockholm, Sweden.

Stevenson, M., & Wilks, Y. (2001). The interaction of knowledge sources in word sense disambiguation. *Computational Linguistics, 27*(3), 321–349.

Strapparava, C., Gliozzo, A., & Giuliano, C. (2004). Pattern abstraction and term similarity for word sense disambiguation: IRST at Senseval-3. *Proceedings of the 3rd International Workshop on the Evaluation of Systems for the Semantic Analysis of Text (SENSEVAL-3)*, Barcelona, Spain (pp. 229–234).

Swinney, D. A. (1979). Lexical access during sentence comprehension: (Re)consideration of context effects. *Journal of Verbal Learning and Verbal Behavior, 18*, 645–659.

Swinney, D. A., & Hakes, D. T. (1976). Effects of prior context upon lexical access during sentence comprehension. *Journal of Verbal Learning and Verbal Behavior, 15*, 681–689.

Tabossi, P. (1988). Accessing lexical ambiguity in different types of sentential contexts. *Journal of Memory and Language, 27*, 324–340.

Taft, M. (1991). *Reading and the mental lexicon*. Hove, East Sussex: Lawrence Erlbaum Associates.

Taft, M., & Forster, K. I. (1975). Lexical storage and retrieval of prefixed words. *Journal of Verbal Learning and Verbal Behavior, 14*, 638–647.

Tanenhaus, M. K., Leiman, J. M., & Seidenberg, M. S. (1979). Evidence for multiple stages in the processing of ambiguous words in syntactic contexts. *Journal of Verbal Learning and Verbal Behavior, 18*, 427–440.

Towell, G., & Voorhees, E. M. (1998). Disambiguating highly ambiguous words. *Computational Linguistics, 24*(1), 125–145.

Veronis, J. (2001). Sense tagging: Does it make sense? *Paper Presented at the Corpus Linguistics 2001 Conference*, Lancaster.

Veronis, J., & Ide, N. (1990). Word sense disambiguation with very large neural networks extracted from machine readable dictionaries. *Proceedings of the 13th International Conference on Computational Linguistics (COLING-90)*, Helsinki, Finland. (pp. 389–394).

Voorhees, E. M. (1999). Natural language processing and information retrieval. In M. T. Pazienza (Ed.), *Information extraction: Toward scalable, adaptable systems*. London: Springer.

Vossen, P., & Copestake, A. (1993). Untangling definition structure into knowledge representation. In E. J. Briscoe, A. Copestake, & V. de Paiva (Eds.), *Inheritance, defaults and the lexicon*. Cambridge: Cambridge University Press.

Vossen, P., Meijs, W., & den Broeder, M. (1989). Meaning and structure in dictionary definitions. In B. Boguraev and E. J. Briscoe (Eds.), *Computational lexicography for natural language processing*. Harlow: Longman.

Wilks, Y. (1975a). An intelligent analyzer and understander of English. *Communications of the ACM, 18*(5), 264–274.

Wilks, Y. (1975b). A preferential, pattern-seeking, semantics for natural language inference. *Artificial Intelligence, 6*, 53–74.

Wilks, Y. (1998). Senses and texts. *Computational Linguistics and Chinese Language Processing, 3*(2), 1–16.

Wilks, Y., & Stevenson, M. (1996). The grammar of sense: Is word-sense tagging much more than part-of-speech tagging? *Technical Report CS-96-05*, University of Sheffield.

Wilks, Y., & Stevenson, M. (1997). Sense tagging: Semantic tagging with a lexicon. *Proceedings of SIGLEX '97 Workshop: Tagging Text with Lexical Semantics: Why, What, and How?*, Washington, DC.

Wilks, Y., & Stevenson, M. (1998). Word sense disambiguation using optimised combinations of knowledge sources. *Proceedings of the 36th Annual Meeting of the Association for Computational Linguistics and 17th International Conference on Computational Linguistics (COLING-ACL '98)*, Montréal, Canada. (pp. 1398–1402).

Wilks, Y., Fass, D., Guo, C., McDonald, J., Plate, T., & Slator, B. (1989). A tractable machine dictionary as a resource for computational semantics. In B. Boguraev & E. J. Briscoe (Eds.), *Computational lexicography for natural language processing*. Harlow: Longman.

Wittgenstein, L. (1958). *Philosophical investigations* (G. E. M. Anscombe Trans.). New Jersey: Blackwell Publishers.

Wu, D., Su, W., & Carpuat, M. (2004). A kernel PCA method for superior word sense disambiguation. *Proceedings of the 42nd Annual Meeting of the Association for Computational Linguistics (ACL 2004)*, Barcelona, Spain.

Yarowsky, D. (1992). Word-sense disambiguation using statistical models of Roget's categories trained on large corpora. *Proceedings of the 14th International Conference on Computational Linguistics (COLING-92)*, Nantes, France (pp. 454–460).

Yarowsky, D. (1993). One sense per collocation. *Proceedings of the ARPA Human Language Technology Workshop*, Princeton, NJ (pp. 266–271).

Yarowsky, D. (1994). Decision lists for lexical ambiguity resolution: Application to accent restoration in Spanish and French. *Proceedings of the 32nd Annual Meeting of the Association for Computational Linguistics (ACL '94)*, Las Cruces, New Mexico (pp. 88–95).

Yarowsky, D. (1995). Unsupervised word sense disambiguation rivaling supervised methods. *Proceedings of the 33rd Annual Meeting of the Association for Computational Linguistics (ACL '95)*, Cambridge, MA (pp. 189–196).

Yarowsky, D. (2000). Hierarchical decision lists for word sense disambiguation. *Computers and the Humanities, 34*(1–2), 179–186.

Yarowsky, D. (2010). Word sense disambiguation. In N. Indurkhya & F. J. Damerau (Eds.), *Handbook of natural language processing*. Boca Raton, FL: Chapman and Hall.

Yarowsky, D., & Florian, R. (2002). Evaluating sense disambiguation across diverse parameter spaces. *Natural Language Engineering, 8*(4), 293–310.

Yore, L. D., & Ollila, L. O. (1985). Cognitive development, sex, and abstractness in grade one word recognition. *Journal of Educational Research, 78*, 242–247.

Yuret, D., & Yatbaz, M. A. (2010). The noisy channel model for unsupervised word sense disambiguation. *Computational Linguistics, 36*(1), 111–127.